Designing and Teaching the Elementary Science Methods Course

What do aspiring and practicing elementary science teacher education faculty need to know as they plan and carry out instruction for future elementary science teachers? This scholarly and practical guide for science teacher educators outlines the theory, principles, and strategies needed, and provides classroom examples anchored to those principles. The theoretical and empirical foundations are supported by scholarship in the field, and the practical examples are derived from activities, lessons, and units field-tested in the authors' elementary science methods courses.

The concept of pedagogical content knowledge (PCK) is used as an organizing framework. PCK describes how teachers transform subject matter knowledge into viable instruction in their discipline. In this case, the construct of PCK incorporates the knowledge science teacher educators need in order to translate what they know about good elementary science teaching into viable science teacher education.

In Part I, chapters on science methods students as learners, the science methods course curriculum, instructional strategies, methods course assessment, and the field experience examples help readers develop their PCK for teaching prospective elementary science teachers. These chapters include special "It's Your Turn" sections to engage the reader. Part II features useful, field-tested "Activities that Work" for putting PCK knowledge into action in the elementary science methods course. "More to Explore" and "Tools for Teaching" sections across the chapters enhance the effectiveness of *Designing and Teaching the Elementary Science Methods Course*—the only professional development guide specifically designed for elementary science methods instructors.

Sandra K. Abell is Curators' Professor of Science Education at the University of Missouri, U.S., where she directs the university's Science Education Center.

Ken Appleton is semi-retired as an Adjunct Associate Professor at Central Queensland University, Australia.

Deborah L. Hanuscin is Assistant Professor of Elementary Education at the University of Missouri, U.S., where she holds a joint appointment in the Department of Physics and Astronomy and Department of Learning, Teaching, and Curriculum.

Teaching and Learning in Science Series
Norman G. Lederman, Series Editor

Abell/Appleton/Hanuscin
Designing and Teaching the Elementary Science Methods Course

Akerson (Ed.)
Interdisciplinary Language Arts and Science Instruction in Elementary Classrooms: Applying Research to Practice

Wickman
Aesthetic Experience in Science Education: Learning and Meaning-Making as Situated Talk and Action

For more information on titles in the Teaching and Learning in Science Series visit **http://www.routledge.com/education**

Designing and Teaching the Elementary Science Methods Course

Sandra K. Abell
University of Missouri

Ken Appleton
Central Queensland University

Deborah L. Hanuscin
University of Missouri

Montante Family Library
D'Youville College

Routledge
Taylor & Francis Group
NEW YORK AND LONDON

First published 2010
by Routledge
270 Madison Avenue, New York, NY 10016

Simultaneously published in the UK
by Routledge
2 Park Square, Milton Park, Abingdon, Oxon OX14 4RN

Routledge is an imprint of the Taylor & Francis Group, an informa business

© 2010 Taylor & Francis

Typeset in Sabon by Wearset Ltd, Boldon, Tyne and Wear
Printed and bound in the United States of America on acid-free paper by Edwards Brothers, Inc.

All rights reserved. No part of this book may be reprinted or reproduced or utilized in any form or by any electronic, mechanical, or other means, now known or hereafter invented, including photocopying and recording, or in any information storage or retrieval system, without permission in writing from the publishers.

Trademark Notice: Product or corporate names may be trademarks or registered trademarks, and are used only for identification and explanation without intent to infringe.

Library of Congress Cataloging in Publication Data
A catalog record has been requested for this book

ISBN10: (hbk) 0-8058-6339-7
ISBN10: (pbk) 0-8058-6340-0
ISBN10: (ebk) 0-203-85913-8

ISBN13: (hbk) 978-0-8058-6339-0
ISBN13: (pbk) 978-0-8058-6340-6
ISBN13: (ebk) 978-0-203-85913-1

Contents

Preface	ix
Acknowledgment	xii
Introduction	1

PART I
Theoretical, Contextual, and Pedagogical Foundations for the Elementary Science Methods Course — 9

1. Perspectives on Science Teacher Learning — 11

   Views about Learning and Teaching 12
   Ideas about Knowledge 13
   Learning Theories 16
   A View of Learning 24
   An Example Application to a Science Methods Course 32
   Conclusion 34
   More to Explore 35
   References 35

2. The Context for Elementary Science Teacher Preparation — 37

   Who is Responsible for Educating Elementary Science Teachers? 37
   The Policy Context: Goals and Standards for Elementary Science Teacher Education 38
   The Program Context: Elementary Teacher Education 40
   The Program Context: Science Courses 44

The Program Context: Field Experience and Partnerships with Local Schools 46
Conclusion 46
More to Explore 47
References 48

3. Orientations to Teaching Science Teachers   50

Introduction 50
What are Orientations to Teaching Teachers? 51
Some Orientations to Teaching Science Teachers 51
The Reflection Orientation and the Science Methods Course 54
Conclusion 59
Tools for Teaching Elementary Science Methods 60
References 63

4. Understanding the Elementary Science Methods Student   65

Introduction 65
Prospective Teachers as Learners 66
Prospective Teachers' Knowledge for Science Teaching 69
Understanding Your Methods Students 74
Conclusion 77
More to Explore 78
References 78

5. Curriculum and Resources for Elementary Science Teacher Education   80

Aims and Goals for the Science Methods Course 81
The Methods Course Curriculum—What to Include? 82
Selecting Resources and Materials 91
Facilities and Equipment 95
Communicating Your Expectations to Students: The Course Syllabus 98
Conclusion 98
Tools for Teaching the Elementary Science Methods Course 100
More to Explore 101
References 101

6. Instructional Strategies for the Elementary Science
   Methods Course 102

   Aspects of PCK Pertinent to this Chapter 103
   Your Orientation to Learning and Teaching 104
   Environment 106
   General Pedagogy 107
   Teaching Models 125
   Teaching How to Plan 133
   Conclusion 135
   More to Explore 136
   References 136

7. Assessment Strategies for the Elementary Methods Course 138

   *Purposes and Examples of Assessment in the Elementary
       Science Methods Course 139*
   Principles of Effective Assessment 144
   Scoring and Grading Assignments 152
   *Putting it All Together: Designing a Methods Course
       Assignment 157*
   Conclusion 161
   Tools for Teaching Elementary Science Methods 162
   More to Explore 168
   References 168

8. Field Experiences in Elementary Science Methods 171

   Benefits of Field Experiences 172
   *Addressing Challenges to the Science Methods Field
       Experience 174*
   Various Models of Science Methods Field Experiences 178
   Student Reflection on the Field Experience 182
   *Conclusion: Design Principles for Science Methods Field
       Experiences 185*
   Tools for Teaching Elementary Science Methods 187
   References 191

PART II
Activities that Work for the Elementary Science Methods Course — 195

ATW 1. Learning about the 5E Learning Cycle: Magnetism	201
ATW 2. Interactive Approach: Floating and Sinking	218
ATW 3. Inquiring into Guided and Open Inquiry: Insect Study	233
ATW 4. Eliciting Student Ideas: The Human Body	247
ATW 5. Using Models and Analogies: Electric Circuits	255
ATW 6. Learning about Discourse: Light and Shadows	270
ATW 7. Integrating Language Arts and Science: A Journey through the Water Cycle	278
ATW 8. Seamless Assessment: The Moon Investigation	287
About the Authors	302
Index	303

Preface

The idea for this book arose from numerous discussions about the knowledge and skills needed by those who teach elementary science methods courses in our respective countries and around the world. Given our collective experience of 60+ years teaching elementary science methods courses, we recognize that teaching prospective teachers entails a repertoire of knowledge about science, science teaching, and teaching teachers. However, individuals who teach the elementary science methods courses sometimes lack experience in one or more of these areas. We wrote this book to provide guidance to faculty members and aspiring faculty members who might need help in designing and enacting instruction in this context.

The book is grounded in the theoretical framework of pedagogical content knowledge (PCK), which describes how teachers transform subject matter knowledge into viable instruction in their discipline. We apply the construct of PCK to the development of science teacher educators. We believe that PCK for teaching teachers integrates the various types of knowledge that science teacher educators need to translate what they know about best practices in elementary science teaching into viable science teacher education. As a starter, the effective elementary science teacher educator needs to have knowledge of science, the nature of science, and elementary science teaching as well as knowledge of general pedagogical principles and the context of instruction. In addition, the elementary science methods instructor needs to understand methods students as learners, the methods course curriculum, and strategies for instruction and assessment in the methods course. These knowledge bases interact with an instructor's orientations to teaching prospective teachers and provide the foundation for course design and enactment.

Those entering the field or currently teaching science teachers typically have had to construct their own ideas for designing and teaching the elementary science methods course from scratch. That's what each of us did early on in our careers. Although some venues exist to share our wisdom of practice through professional organizations and journals, little shared formal knowledge about teaching future elementary science teachers exists in a comprehensive form. In this book, we provide an accessible foundation for your development of PCK for teaching elementary science teachers that is grounded in research and practice. We hope that you will find *Designing and Teaching the Elementary Science Methods Course* to be a practical, yet scholarly guide to assist you in your science teacher education endeavors.

Overview of the Book

Designing and Teaching the Elementary Science Methods Course is divided into two main parts. In Part I, "Theoretical, Contextual, and Pedagogical Foundations for the Elementary Science Methods Course," we use the teacher knowledge framework to discuss the knowledge for teaching teachers needed by science teacher educators. We first discuss the underlying learning theories, policy and program contexts, and orientations to teaching teachers that can guide the teaching of future elementary science teachers.

Our vision of science teacher education is governed, in part, by our epistemological views of teacher learning and knowledge. We explain those views in Chapter 1, "Views about Learning and Teaching." This chapter will help you develop your general pedagogical knowledge for teaching teachers (see Figure I.1). Chapter 2, "The Context for Elementary Science Teacher Preparation," focuses on developing knowledge of context for teaching teachers. We discuss important context variables that affect our work with future elementary teachers of science. In particular, we examine the various stakeholders in elementary teacher preparation—our scientist and education colleagues at the university, our school-based colleagues, and the policy makers that govern teacher education and licensure. In Chapter 3, "Orientations to Teaching Science Teachers," we present a number of possible orientations to designing and teaching the elementary science method course that provide filters through which instructors apply various facets of PCK for teaching teachers. We also provide an in-depth account of one such orientation—the reflection orientation—and discuss the purposes, goals, and beliefs that frame it.

In Chapters 4–8, we examine the important pedagogical considerations for designing and implementing an elementary science methods course. In particular, we attempt to develop an instructor's PCK for teaching elementary science teachers in terms of knowledge of learners (Chapter 4), curriculum (Chapter 5), instruction (Chapter 6), and assessment (Chapter 7) for the elementary science methods course. These chapters provide the nuts and bolts of how to design and teach such a course, using examples from our own teaching. In Chapter 8, we present various models for structuring field experiences that could be partnered with the science methods course and provide a different setting for student learning. The field experience demands the integration of knowledge for teaching teachers in new settings.

We designed the chapters in Part I to represent a conversation with you, the elementary science methods course instructor. Each chapter is framed by a vignette derived from our experience in which we address a problem of practice. We intersperse our ideas and ideas from the literature with opportunities for you to try some things in your own course, in the spirit of learning by doing. We have structured the chapters purposefully to promote your learning as a science teacher educator, by including special sections called "It's Your Turn," "More to Explore," and "Tools for Teaching." We hope these chapters will help you add important ideas to your methods teaching repertoire, including PCK for learners, curriculum, instruction, and assessment. However, we also hope that you will use the chapters flexibly adapting the ideas to meet the needs, interests, and abilities of your students. It is in that spirit that we share our ideas.

We believe that developing expertise in teaching future science teachers rests on knowledge of both principles and cases. We designed the first part of the book to provide the principles and some brief examples of elementary science methods instruction. In Part II, "Activities that Work for the Elementary Science Methods Course," we describe field-tested examples from our experiences instructing science methods courses in the elementary science teacher preparation program. These cases of teaching elementary science teachers include lessons, resources, and tools for your methods course. We call these cases "Activities that Work for the elementary methods course," grounded in our guiding framework of PCK. These "Activities that Work" can be thought of as a beginning repertoire for you as a science teacher educator to adopt or adapt. We have included activities that simultaneously model pedagogical principles (such as routines, techniques, strategies, and models for instruction and assessment) for future elementary teachers, illustrate nature of science aspects, and help prospective teachers develop understandings of science concepts for different elementary grade ranges (within the disciplines of life, physical, and earth science).

Learning to teach future science teachers is a lifelong endeavor. Our teaching of methods courses and our research about teacher learning have helped us to form our ideas about designing and teaching the elementary methods course. We hope that you, as a teacher of future elementary science teachers, can benefit from our wisdom of our practice as you put these ideas into action in your own teaching. Of course, your teaching will also be influenced by the interest and enthusiasm you bring to the task. We encourage you to nurture your passion as you build your knowledge for teaching the elementary science methods course.

S.K.A.
K.A.
D.L.H.

Acknowledgment

The authors acknowledge Emily M. Walter, doctoral student in Science Education at the University of Missouri, for her contributions to this book. Her meticulous work in checking references, gathering permissions, correcting our grammar, and formatting the writing has been invaluable. The book is much stronger for her careful and thoughtful work. Thanks, Emily!

Introduction

Teacher expertise is one of the most important factors in determining student achievement. Put another way, teacher knowledge—what teachers know and can do—affects what their students learn. Similarly, what teacher educators know and can do affects what their students, prospective teachers, will learn. Thus the role of teacher education, in particular the methods course, is critical to developing high quality teachers. Science teacher educators need support in building the knowledge necessary to teach the science methods course. This book is intended to help science teacher educators build their knowledge of teaching teachers so they can facilitate the development of expertise in science teaching among prospective elementary[1] school teachers. This expertise will in turn allow elementary teachers to help their students learn science.

As science educators, we have been in the business of preparing individuals to teach elementary science for a number of years. During more than 19 years of combined experience as elementary classroom teachers, we shaped our views of science teaching and learning. Furthermore, our collective experience of 60+ years in science teacher education helped us develop a theoretical foundation and pedagogical repertoire for teaching science teachers. Our work in various countries, mainly the U.S. and Australia, but also Canada, Honduras, New Zealand, Portugal, Thailand, Taiwan, and the U.K., and in several different institutions of higher education, influenced our understanding of the role of context and policy in science teacher education. Furthermore, our research on science teacher learning influenced our perspectives on designing and teaching the methods course. We offer this book as a synthesis of our knowledge (of practice, policy, and research) with the hope that you and your teacher education students will benefit.

---

1. We use the term "elementary" to refer to teachers of children, ages 5–12, recognizing that in some parts of the world, the term "primary" typically is used.

Purpose of the Book

This book is unique in that it is written expressly for instructors of elementary science methods courses, not for their students. The book is meant to communicate with science teacher educators around the globe. Although our contexts and challenges may differ, our goals do not. We all aim to help future teachers teach science for understanding. We try to improve science teaching and learning in schools by preparing teachers with expertise to support science learning. We hope to generate a cadre of individuals who have the knowledge, skills, and dispositions to be successful teachers of science. However, instructors of elementary science methods courses themselves differ in their expertise and experience in teaching science in the elementary school and in teaching teachers. For example, some methods course instructors have never taught in K-12 settings, others have taught science only in high schools, and others have substantial experience teaching science in the elementary school. While some methods instructors have vast amounts of experience apprenticing as science teacher educators before teaching their own methods course, each instructor faces the task of teaching his/her first solo methods course at some point in time. The purpose of this volume is to provide a theoretical framework and practical examples for science teacher educators to adopt and adapt as they plan and carry out instruction in elementary teacher preparation programs.

Our Guiding Framework: Science Teacher Knowledge

We believe that the main aim of an elementary science methods course is to help prospective teachers start to build the knowledge that will enable them to teach science effectively. Over 20 years ago, Shulman (1986, 1987) defined a set of knowledge bases that teachers draw on when they teach. Science teachers know science, they know about teaching and learning in general, and they understand the context of their teaching situation. More importantly, according to Shulman, science teachers have a special kind of knowledge for teaching science that includes knowledge of science learners, curriculum, instruction, and assessment. Shulman called this set of knowledge, pedagogical content knowledge, or PCK for short. We believe that PCK for learners, curriculum, instruction, and assessment is filtered through a teacher's orientation to science teaching as it is put into action (Abell, 2007; Grossman, 1990; Magnusson, Krajcik, & Borko, 1999). It is this specialized knowledge—PCK—that characterizes the work of science teachers. It is this specialized knowledge—PCK—which we aim to help our future teachers develop.

Shulman's notion of PCK has been refined and revised by a number of science educators over the years (e.g., Barnett & Hodson, 2001; Carlsen,

1991; Gess-Newsome & Lederman, 1999; Loughran, Mulhall, & Berry, 2004; Magnusson et al., 1999; Veal & Kubasko, 2003). Appleton (2006) presented a view of PCK specific to the elementary science teacher. From his perspective, inputs to PCK include science content knowledge and general pedagogical knowledge, as well as science teaching specifics such as:

- orientations to teaching and learning science;
- confidence to teach science;
- knowledge of different aspects of the teaching context, such as curriculum, assessment, instructional strategies, and resources;
- knowledge of students.

According to Appleton (2003a), when planning and teaching science, the elementary teacher draws on these various types of knowledge to generate a new form of knowledge, science PCK. PCK is generated and regenerated by the acts of planning, teaching, and reflecting; and is usually communicated to others as descriptions of activities in which the students are engaged. Appleton (2003b) called these "activities that work," employing an expression that elementary teachers themselves used.

We contend that a parallel model of teacher knowledge exists for those who teach future teachers—science teacher educators (Abell, Park Rogers, Hanuscin, Gagnon, & Lee, 2009). Figure I.1 shows the types of knowledge needed for teaching science teachers, based on Grossman's (1990) model.

- *Subject Matter Knowledge.* In the case of teaching teachers, the subject matter knowledge that an elementary science teacher educator needs includes both science content knowledge and knowledge for teaching elementary science. Elementary methods course instructors have built this knowledge through their experiences as science learners and teachers.
- *Pedagogical Knowledge.* Science teacher educators need general pedagogical knowledge for university-level teaching and learning. For example, they need to know what information should be part of a syllabus, how to deal with problematic classroom situations, and how to assign grades.
- *Knowledge of Context.* Because preparing future teachers is a collaborative effort among scientists, educators, school personnel, and policy makers, methods course instructors need to understand the context of teacher education in their country, state, and university. Thus it is important to understand the requirements for becoming a teacher in terms of federal/state standards and the structure of the university teacher education program.

These three types of knowledge are supported and transformed by the science teacher educator's PCK for teaching teachers in the science methods course (see Figure I.2). PCK for teaching science teachers includes knowledge about curriculum, instruction, and assessment for teaching science methods courses and supervising field experiences, as well as knowledge about teacher learning. Each of these knowledge bases is filtered through a teacher educator's orientation to teaching science teachers. Specifically, the elementary science methods instructor needs:

- *Orientation to Teaching Teachers.* A science teacher educator's orientation to teaching the elementary methods course includes both an orientation to teaching science and an orientation to teaching science teachers. These orientations take into account the science educator's views of science teaching and learning and teacher learning. Orientations act as a lens for other kinds of PCK.
- *PCK for Learners.* A science teacher educator's knowledge of elementary methods students includes knowing general characteristics of students' development, common misconceptions held by students about science and science teaching, and characteristics specific to a particular group of methods students. The science teacher educator needs to understand how elementary methods students come to learn science and science teaching, and what kinds of difficulties (both affective and cognitive) they will have in developing their PCK for teaching elementary science.
- *PCK for Curriculum.* The methods course instructor needs to know appropriate curricular goals elementary methods course. Course goals will be based on certification standards and goals of the teacher preparation program, in addition to the instructor's specific goals for becoming an elementary science teacher. The instructor also needs knowledge of resources for teaching the elementary methods course, including texts, online resources, and other curriculum materials.
- *PCK for Instructional Strategies.* Science teacher educators need to know a variety of instructional strategies that are effective for teaching various concepts and techniques about teaching and learning elementary science. They also know particular activities and representations that work in the methods course.
- *PCK for Assessment.* The elementary science teacher educator needs knowledge of assessment appropriate for the elementary science methods course, including formative and summative assessment strategies and example assignments. The teacher educator also needs knowledge of how to use assessment data to inform instruction.

In this book, we provide a road map for the various types of knowledge you will need for teaching future elementary teachers in the methods course setting based on our model of PCK for teaching science teachers.

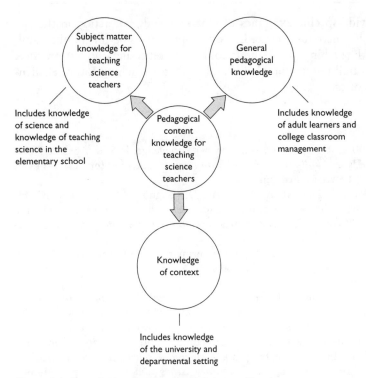

Figure 1.1 Knowledge for Teaching Science Teachers.

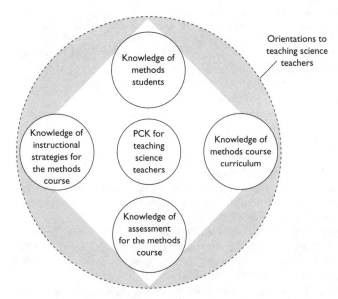

Figure 1.2 Science Teacher Educator PCK for Teaching Prospective Teachers.

We also provide specific examples of that knowledge in action in the elementary science methods course. It is our hope that using this book, simultaneously with teaching the methods course and reflecting on your practice, will help you build your knowledge and skills for teaching future elementary teachers of science.

References

Abell, S. K. (2007). Research on science teacher knowledge. In S. K. Abell & N. G. Lederman (Eds.), *Handbook of research on science education* (pp. 1105–1149). Mahwah, NJ: Lawrence Erlbaum.

Abell, S. K., Park Rogers, M. A., Hanuscin, D., Gagnon, M. J., & Lee, M. H. (2009). Preparing the next generation of science teacher educators: A model for developing PCK for teaching science teachers. *Journal of Science Teacher Education, 20,* 77–93.

Appleton, K. (2003a). How do beginning primary school teachers cope with science? Toward an understanding of science teaching practice. *Research in Science Education, 33,* 1–25.

Appleton, K. (2003b, July). *Pathways in professional development in primary science: Extending science PCK.* Paper presented at the Annual conference of the Australasian Science Education Research Association, Melbourne, Australia.

Appleton, K. (2006). Science pedagogical content knowledge and elementary school teachers. In K. Appleton (Ed.), *Elementary science teacher education: International perspectives on contemporary issues and practice* (pp. 31–54). Mahwah, NJ: Lawrence Erlbaum in association with the Association for Science Teacher Education.

Barnett, J., & Hodson, D. (2001). Pedagogical context knowledge: Toward a fuller understanding of what good science teachers know. *Science Education, 85,* 426–453.

Carlsen, W. S. (1991). Subject-matter knowledge and science teaching: A pragmatic perspective. In J. Brophy (Ed.), *Advances in research on teaching: Vol. 2. Teachers' knowledge of subject matter as it relates to their teaching practice* (pp. 115–144). Greenwich, CT: JAI Press.

Gess-Newsome, J., & Lederman, N. G. (Eds.). (1999). *Examining pedagogical content knowledge: The construct and its implications for science education* (pp. 51–94). Boston: Kluwer.

Grossman, P. L. (1990). *The making of a teacher: Teacher knowledge and teacher education.* New York: Teachers College Press.

Loughran, J., Mulhall, P., & Berry, A. (2004). In search of pedagogical knowledge in science: Developing ways of articulating and documenting professional practice. *Journal of Research in Science Teaching, 41,* 370–391.

Magnusson, S., Krajcik, J. S., & Borko, H. (1999). Nature, sources, and development of pedagogical content knowledge for science teaching. In J. Gess-Newsome & N. G. Lederman (Eds.), *Examining pedagogical content knowledge* (pp. 95–132). Dordrecht, the Netherlands: Kluwer.

Shulman, L. S. (1986). Those who understand: Knowledge growth in teaching. *Educational Researcher, 15*(2), 4–14.

Shulman, L. S. (1987). Knowledge and teaching: Foundations of the new reform. *Harvard Educational Review, 57*, 1–22.

Veal, W. R., & Kubasko, W. R. (2003). Biology and geology teachers' domain-specific pedagogical content knowledge of evolution. *Journal of Curriculum and Supervision, 18*, 344–352.

Part I

Theoretical, Contextual, and Pedagogical Foundations for the Elementary Science Methods Course

Chapter 1

Perspectives on Science Teacher Learning

> Imagine, if you will, an informal conversation in the college coffee shop. The elementary science methods instructors, Noeline, Sam, and Joan have met with two colleagues, Asmar and Peter, from the science department. They are all seeking relief from marking assignments and exams. Peter complains, "I get so frustrated with these students! They get the simplest things wrong." Asmar nods in agreement, "Doesn't matter how many times you tell them, they still mess it up." Peter continues, "I mean, I've told them three times, and it's in their text book. Why can't they just learn it?" They are both looking at Noeline as if she can explain the mystery. What could she say?

Do Asmar's and Peter's comments sound familiar and align with your experience? Is there an explanation? In this chapter, we explore some possible answers, and ways they might be addressed, by exploring ideas about how your prospective teachers learn.

It's Your Turn

In the above scenario, Asmar and Peter have revealed aspects of their views about learning.

What views about learning do Peter and Asmar seem to have?
How would you answer if you were Noeline?
Could you make any suggestions as to how Peter and Asmar might change the way they do things?

Views about Learning and Teaching

When you, the elementary science methods instructor, design and implement a science methods course, you draw upon your own pedagogical content knowledge (PCK) for teaching teachers. Your orientation to teaching and learning (see Chapter 3) has influenced the development of your PCK for teaching teachers, and also influences the types of instructional strategies that you use. Furthermore, your orientation to teaching and learning has two facets: an orientation to teaching and learning science, and an orientation to teaching and learning in science teacher education. Your views about the former need to be consistent with those you are promoting in your science methods course, and your views about the latter also need to be consistent with that view, but also need to take into account aspects of adult learning, and the particular characteristics of prospective elementary teachers.

A key goal of your methods course, explicitly stated or implicit, is to help your methods students develop sufficient PCK for teaching science to enable them to commence teaching science—what might be called a "starter pack." So, to design and implement a science methods course, you need to have a clear idea about the development of PCK for teaching elementary science and your PCK for teaching teachers. In this chapter, we examine several ideas about learning that are pertinent to your views of teaching and learning in science teacher education, and invite you to reexamine your own views in the light of what you discover.

Our Interpretation of the Scenario

If we read between the lines, Peter and Asmar seem to think that knowledge can be put into students' heads if it is repeated enough, and that the knowledge can then be relayed back to him, as delivered, in an assessment exercise. As far as Peter is concerned, learning is a simple exercise, where knowledge is somehow crammed into a person's head, analogous to an empty glass being filled with water from a jug.

This is a fairly common view held by many college and university professors, some teachers, most students, and the majority of those in the community. Unfortunately, it is not a particularly accurate view according to research, and it can lead to ineffective teaching, inappropriate curriculum, and poor assessment. At the moment, we have no inkling of what views Noeline holds. What she believes about learning is going to be passed on to her students during her methods course. Could she be passing on a view that perpetuates a defective idea, or would she have a different view? As you read the next section, compare the ideas to your own, and the ways in which you may have considered answering Peter and Asmar.

Views about teaching are partially derived from views about how people learn, which in turn are influenced by epistemological views (views about how knowledge is organized and developed). So, in this part of the chapter, we focus on some contemporary views about knowledge and learning, and how these relate to teaching.

Ideas about Knowledge

A major difficulty in developing views about knowledge is that we are speculating about what goes on in people's heads. Brain scientists and neurologists have worked out which parts of the brain have particular roles, and can map brain activity when certain tasks are undertaken. We know that neurons (nerves in the brain) can form many interconnections with other neurons, and that the more frequently connections are used, the "stronger" they become; just as a path can be worn in grass by many feet passing the same way. But this is only telling us about the physical. While it gives us clues about how knowledge is arranged and stored in the brain, how it is retrieved, and how it is added to, the picture is very incomplete. So we must rely on some theories about this, which are consistent with both the physical evidence, and studies of human learning. We cannot pretend that this book is about psychology, so we will only refer to those theories that we think are contemporary, have currency in science education, and are helpful to science methods instructors. Schema theory gives us one way of thinking about knowledge organization and extension (epistemology) that is very useful.

Schema Theory

The key idea of this theory is that our past experiences are linked mentally in our memories, to form an organized structure of ideas, which psychologists call a cognitive structure. It is postulated that experiences are stored in the memory as separate events, but those having some commonality are linked in clusters. These clusters form an idea, consisting of multiple links across a network of experiences. Similarly, ideas with commonalities are linked to each other forming a complex network of linked ideas and experiences. These are often represented diagrammatically by Venn diagrams or concept maps.

A network of mental ideas centered round some commonality is called a schema (the plural of this Latin word is schemata). Often the anglicized version of the word, scheme (plural, schemes), is used. An example of a schema would be a simple concept like "duck." Imagine, if you will, a young child walking past a pond with a parent, pointing to a white bird swimming on the surface, and asking, "What's that?" The parent's answer, "A duck," allows the child to store in memory the word and the visual impression of a swimming white creature. They later encounter a white swan on the pond, and the child points, exclaiming "Duck!" The parent responds, "No dear, that is a swan. See, it has a long neck, and is a lot bigger than the duck." The child processes this information, forming in her mind links between this experience and the earlier memory of "duck," specifically linking to remembered aspects of neck length and body size. They later encounter a Mallard duck, and again there is an exchange where the parent patiently points out the differences and similarities between this duck, the white duck and the swan. In this way, the child builds a network of memories of different water birds and their language labels, with specific differences and similarities noted. A schema of "duck" is being constructed, with links to other schemata centered round "swan," and perhaps later, "goose." Each of these schemata is in turn linked to the others to form a general schema of water birds, and more generally, one of birds.

Notice in the above example, the notion that schemata are constructed. This is an important idea when we come to look at learning theories. Construction is a subconscious activity, but it does require deliberate attention to experiences and information extracted from them. Also, note that networks between schemata are constructed, linking those with similarities, and often forming more general schemata. Network construction is also a subconscious activity, which still requires attention to experiences, remembered experiences, and information embedded in them. Finally, note that new information gathered from experiences via the senses is incorporated into memory by association with existing schemata.

This process of making links between experiences and ideas can be considered as analogous to neurons in the brain forming links to other neurons—though perhaps in very different parts of the brain. According to this view, then, at some point soon after neuron development in a baby begins, the neurons start forming links with other neurons. Some links endure; some fade. As sensory organs begin to deliver information to the brain (such as sound penetrating into the womb), simple schemata begin to develop. As the brain receives new information the schemata are modified and extended. This process continues as the child grows. There is evidence to suggest that neuron growth (or at least replacement) continues throughout life, and that there are a couple of periods of rapid neuron growth in children.

Schema theory also provides a perspective on what happens when we remember something. A sensory stimulus that occurs in a particular context is linked to a schema selected by the brain. This both helps us interpret the stimulus, and activates a string of memories related to that schema: those memories have been recalled. Problems in recall can occur if the brain selects a different schema that activates a different set of memories, or if the link to a particular memory has faded.

Revisiting the Scenario

Noeline knows about schema theory, and could use it to explain why Peter and Asmar's students are having difficulty with their courses. Her thinking goes something like this:

> During a lecture the students are being presented with a lot of new information all at once, which they have to fit into their existing schemata if they are to learn it. The trouble is, a lot of the ideas are abstract, and the students have few schemata that can form links to the new information. Because so much is coming so fast to just a few schemata, some links become established, but some are too weak.
>
> Some information is simply missed because the students are concentrating in an effort to fit earlier information into schemata, or to write notes. More information is provided, and to comprehend this new information, links must be made to information only just processed. A rather tenuous string of links may be formed, with few if any cross-links to other schemata. Such a tenuous string will not readily survive unless revisited in the near future: revisiting reinforces the tenuous links, and helps form cross-links to other schemata.
>
> In some cases, inappropriate schema may have been activated in the first place, so that the tenuous string being tentatively constructed makes no sense. It is even possible that the schema originally activated includes some broken or inappropriate links on which the newly forming links are dependent. They too are destined to fade unless the whole schema can be reconstructed.

It's Your Turn

How would you use the above thoughts to explain Noeline's perspective to Peter and Asmar?

If you wish to explore other theories about epistemology, you will need to consult other books, particularly those from the field of cognitive psychology. An excellent account of one version of schema theory from an educational perspective is provided by Guy Claxton (1990). Please note that we have made a deliberate choice to discuss schema theory, and not to discuss other theories.

Other Implications for Elementary Science Teacher Education

From this epistemological view and research into beginning elementary teachers' characteristics with respect to science, we suggest that prospective teachers bring with them to their science methods course existing schemata about the nature of learning and the nature of teaching, formed from their own schooling experiences. As well, many prospective elementary teachers have experienced difficulty with their own learning in science, and therefore associate with science a learned dislike or even fear, and a perception that they are incapable of learning it (for more ideas about the elementary methods student, see Chapter 4). Mere information provision of science content using traditional strategies that turned them off science in the first place, will therefore not benefit these future teachers. Totally new schemata for science and for science teaching need to be developed, with associated feelings of success and enjoyment, for novice elementary teachers to be equipped to begin teaching science. How the elementary science methods instructor may go about doing this is outlined in later chapters.

Learning Theories

While epistemology deals with how knowledge is organized and acquired, many in science education have focused more on the learning process, especially in formal educational settings. As you will see, there are some very different viewpoints about learning. In our scenario, Asmar and Peter, for instance, seem to think that learning is a process of students somehow putting what they are told into their heads. They could even adapt their thinking to schema theory—just replace schema acquired informally and intuitively with the ones provided by experts like themselves.

> ### It's Your Turn
>
> You probably have some ideas about learning that you have picked up from various places. On a notepad jot down some dot points that you think might be important ideas in explaining how learning occurs.

Depending on factors like when you did your university course, the interests of your university professors, and your own interests, the ideas that you have about learning could come from a number of theories that have held sway over the years. Most educational psychology books provide a comprehensive overview of these ideas, so we will emphasize just a few contemporary theories that are influencing thinking about how to teach science. Compare the ideas in your list with those below.

Overview of the Main Theories about Learning

One of the earlier theories that had considerable influence on science education was behavioral learning theory. This focused on the desired learning outcomes, the consequent inputs that would generate the desired learned behavior, and the rewards or punishments that would encourage the learned behavior. The theory was based on Pavlov's work with getting dogs to salivate at the sound of a bell by providing appropriate rewards, and other similar experiments. In applying these ideas to education, learning tasks were broken into small, achievable tasks that fitted with a hierarchy of learning goals. The idea was that on achieving each goal, the student would be rewarded by success associated with the achievement. Students were tested on each set of goals that they had to satisfy before progressing. Educationalists such as Gagné were heavily involved during the 1960s and 1970s in developing the main elementary science program based on these ideas: Science—A Process Approach.

Since the late 1970s, most science educators have thought that behavioral theories are only partially useful, and have sought other ideas. Behavioral theories are now thought to be mainly useful when teaching lower-level cognitive practices and skills so we will devote no further time to this.

> It's Your Turn
>
> *What aspects of behavioral theory align with your thinking? Is this an area that you would like to find out more about?*

Social Cognitive Theory

An offshoot of behavioral theory, Social Cognitive Theory (Bandura, 1986), is still considered useful today, especially in teacher education, so we provide a brief summary. A key component of Bandura's theory is social modeling, where learners acquire new behaviors by observing and copying others. The extent that modeled behaviors are practiced and exhibited by a learner depend on whether there are perceived rewards or punishments associated with copying the behavior. A young child who watched a violent cartoon, for example, may act violently toward other children if he perceives a "reward" of having greater access to toys.

Another idea associated with exhibiting modeled behavior is that environmental situations cue the behavior to occur. A well-known classroom example is the raising of hands when a teacher asks a question. The cue is the question, eliciting the behavior of raising hands. However, each child will choose whether or not to raise her hand depending on what she expects will happen. If she has an answer that she is certain is correct, she will raise her hand enthusiastically because she expects a reward of public affirmation of her academic ability. If she is uncertain of the answer, she may raise her hand to show others that she knows the answer (a reward), but try to be less obvious—being selected and giving an incorrect answer constitutes a form of punishment. Alternatively, she may choose not to raise her hand: the expectation that she will be wrong is too great, and the minor punishment of not raising her hand is preferable to the greater one of possible public humiliation arising from giving a wrong answer.

Bandura's theory also emphasizes the importance of learners attending to, and perceiving, the salient features of the modeled behavior. This means that who is perceived as a model is important, for instance, whether the model has status or power. Learner characteristics are also important. The notion of self-efficacy (Bandura, 1997), perceived competence in a particular behavior, has found particular application in science teaching: elementary teachers tend to feel less competent in teaching science than other subjects, and are said to have low self-efficacy. The significance of the behavior being modeled to the learner is also a factor that influences the extent that the behavior may be attended to and copied. Further, each of student perceptions of good models, learner characteristics, and perceived significance of the modeled behavior affect motivation.

> **It's Your Turn**
>
> *What is there about Social Cognitive Theory that strikes a chord with you?*
> *Were any of these points on your list?*
> *How could you apply this to your science methods class?*

An important application of this theory to a science methods class is that the role of the instructor (the prospective teachers' science education social model) is critical in helping prospective teachers overcome their fears of science, and see it in a positive light (Olson & Appleton, 2006).

Constructivist Theories

Indisputably, the main learning theories that have influenced science education since the early 1980s are constructivist theories. There is not one constructivist theory, but several, each with a different perspective and emphasis. The theories can be organized into one of three groups: personal constructivism, social constructivism, and information processing (McInerney & McInerney, 2006).

> **It's Your Turn**
>
> Given that constructivism has been around for over 20 years, you have probably come across it before.
>
> *Were some aspects of your list about learning related to constructivism?*
>
> Perhaps this introduction has triggered a few more ideas. Add them to your list.

In the following paragraphs, we briefly describe some of the main ideas in these three different groups of theories. More detail can be obtained from the references or other books such as educational psychology books. Then we provide a view of learning based on many of these theories and through further research.

Personal Constructivism. There is no one theory that fits under this general term. Rather, there are a number of views collectively labeled personal constructivism, or cognitive constructivism, as it is often called. These have been proposed by authors spanning a number of fields such as cognitive psychology and education over many decades. A common feature of cognitive constructivist views is that a learner uses past experiences to interpret and understand new experiences. That is, both new information available to a learner, and what the learner already knows, determine what new information is added to a learner's knowledge repertoire.

The most important theorist to education was Piaget (1974, 1978). While many educationalists have focused on aspects of his theories that emphasize developmental stages, we prefer to focus on his ideas about how a learner fits new knowledge into existing schemata. We provide just a brief outline of the essentials of his ideas. Two key ideas were assimilation of new knowledge into existing knowledge, and accommodation, where existing structures are modified to fit new knowledge. A common term for this process in the constructivist literature is cognitive change.

According to Piaget, assimilation is the process whereby a person uses existing mental schemata to interpret sensory input, so the person can make sense of the input (Furth, 1969). In a cognitive sense, the sensory input is given meaning by using existing mental schemata that interpret the input in terms of what is already known. Piaget insisted that assimilation is an active structuring of experience, and not merely an internal copying of some external reality (Piaget, 1974). Accommodation is the process by which a person adjusts internal schemata to the assimilated data; it is the person's response to assimilated experience (Furth, 1969). In a cognitive sense, if there is a mismatch between the internal schemata and the interpretation of experience, the internal schemata are adjusted to allow the newly assimilated data to fit. Such a mismatch Piaget called disequilibrium, which contrasts with the hypothetical state of equilibrium where a person is not experiencing a mental perturbation. The underlying issue for Piaget was that a person actively constructs meaning from experience according to internal schemata, and is in turn affected by the experience either by a change to internal schemata, or by some exploratory action, or both (Piaget, 1970). It should be noted that Piaget included in his discussion of cognitive schema and systems, not only concepts, but the operations necessary to transform experience and concepts (Glasersfeld, 1987; Piaget, 1974; Vuyk, 1981).

The Piagetian notion of disequilibrium, in particular, has had considerable influence on the science education literature on learning and teaching. Piaget suggested that changes in cognitive structures result from "disturbances" to the assimilation process, which initiates compensatory behaviors (Piaget, 1978, p. 82). Also noteworthy is that he linked affective responses to this:

> In our interpretation of the connections between any cognitive construction and outside disturbances with their resultant compensating reactions ... it goes without saying that an essential place should be reserved for need and consequently for interests. On the one hand, interest is the motivating force or value in any assimilation scheme ... On the other hand, need is the expression of a schema's momentary non-functioning, and, from the cognitive viewpoint, it thus corresponds to a gap or a deficit.
>
> (Piaget, 1978, p. 83)

Other people associated with this group of theories include Kelly (1955, 1963) and Claxton (1984, 1990).

Social Constructivism. Social constructivist ideas form another cluster of theories that emphasize the role of the social and cultural context in learning, and the mediating roles played by others. Many of these theories are derived from Vygotsky's (1962, 1978) work, a person who was a contemporary of Piaget. However, his writings were in Russian, the political circumstances in his home country (Stalinist U.S.S.R.) were difficult, and the cold war made communication outside communist states difficult; so western theorists initially paid scant attention to his ideas. Social constructivist ideas have since become fairly dominant in constructivist thought. While there are many important considerations about Vygotsky's theory, there is one important idea on which we elaborate just a little because of its importance to current thinking, and because it is an assumed feature of the learning view outlined later. This is the Zone of Proximal Development (ZPD).

For Vygotsky, the ZPD represented the potential conceptual development by a child in a particular area. We can better get a grasp of this idea through an example. Consider a child observing a caterpillar moving across a leaf. The child may watch it for a minute or two, and may be able to describe the movement as walking or crawling, drawing on words and concepts already available. So the child has extended its understanding in that caterpillars are also considered to be able to walk or crawl. In comparison, consider the following interaction between an adult and a 5-year old watching the caterpillar:

ADULT: What's it doing Michael?
MICHAEL: Crawling.
ADULT: It's moving all right. Have a closer look. See how it's moving its front forward. Now what's it doing?
MICHAEL: Humping up its back.
ADULT: It is, isn't it? Now, see how it moves its back part up when it humps its back?
MICHAEL: It's going out then humpy [the child gestures with his hands, showing how the front part moves forward, stretching the caterpillar out, then humps its back to bring its rear end forward].

The conversation continued, with the adult asking Michael where its feet were. Note that it was only after the adult asked Michael to look again, and pointed out what to look at, that he was able to understand in more detail the movement of the caterpillar. The potential was there for Michael to grasp this idea, but he was only able to reach that potential because of interaction with someone else who helped him. This area of potential achievement is what Vygotsky thought of as the ZPD. The idea highlights the importance of social interaction with a more capable person to help students reach greater levels of achievement than they could by themselves.

Vygotsky's ZPD has led to the development of some new educational jargon: scaffolding (Hodgson & Hodgson, 1998a, 1998b). This refers to the actions (including verbal ones) that a teacher needs to implement to help students achieve greater understanding. This involves knowing the student well enough to be able to assist including the student's existing capability and level of understanding in what is to be taught; knowing the level of difficulty of the task/idea being taught; and being able to provide a series of specific interventions (teaching actions and/or verbal ones) that help the student progress from what they already know to the predetermined goal. This is not a one-off event (as often suggested in the use of the jargon), but may need to be repeated a number of times, preferably in different ways. However, each time a part of the scaffolding sequence needs to be removed so that the student is pushed toward independence rather than continued reliance on the teacher. This is analogous to a metal scaffold erected around a building under construction—it is there only as long as needed, and is removed progressively as parts are completed.

An extension of Vygotsky's work is that learning takes place in social situations, leading to the term "situated cognition." Learning is seen as being socially mediated, and bound to the context and specific situation in which it occurs. That is, it is impossible to separate what is being learned from the context where the learning is occurring (e.g., Greeno, 1997). Because this theory emphasizes the role of enculturation (Rogoff, 1990), it has become important in teacher education, particularly influencing ideas about the practicum.

Information Processing. The final constructivist view of learning that we summarize comes from comparisons of human learning with the functions of computers and artificial intelligence (e.g., Flavell, 1985), hence the term information processing. A common characteristic of these views is the use of diagrams to represent ideas. Important notions from computing include information input, information processing, storing information, and retrieving information. The focus is therefore on what occurs inside a learner's head during learning, and is clearly related to aspects of personal constructivism. According to McInerney and McInerney (2006) there are four main foci:

- *Information input*, where the learner receives information through the senses and routes it to the appropriate part of the brain. This is analogous to the various data input devices used with computers, such as keyboard, and mouse. Different inputs are attended to, depending on expectations and interests. For instance, pubescent boys may select to attend to input from an attractive female rather than from the teacher because their expectations have shifted from a learning focus to a sexual one. Alternatively, a learner's expectation of a science activity may be to be the first to finish, in comparison to the teacher's expectation that a key science idea be identified.
- *Information processing* refers to the sensory input being placed in short-term memory, and links made to existing memories for interpretation. This is analogous to a computer's random access memory (RAM). Too much information at once can result in cognitive overload. For instance, if the color-code key to a diagram is placed outside of the diagram instead of inside it, the consequent greater processing required prevents many young children from making sense of the diagram. The science misconceptions research also showed how the memories retrieved for interpreting the information determine how the information is interpreted. For instance, if learners hold a view of electric circuits similar to the "clashing current" model (Osborne & Freyberg, 1985), where current is believed to flow from both ends of the battery and "clash" in the bulb to make it light, then there is potential for another activity involving circuits to be interpreted differently from how the teacher may think of it.

- *Storing information* refers to the sensory input and its interpretation being stored permanently in the brain, with links to other related experiences. Emotional states associated with the experience are also stored. This is analogous to saving computer files to hard drives, CDs/DVDs, or compact flash drives. The effectiveness with which material is placed into long-term memory is in part related to the importance that the learner places on the information. Nuthall (1999, 2001) found that for information to be placed into long-term memory and later retrieved, students need at least three separate experiences with the same information, each experience no longer than 2 days after the previous one.
- *Retrieving information* refers to how, at a later time, the remembered information is retrieved from long-term memory to help interpret another experience, or to demonstrate learning. This is analogous to accessing a file on the hard drive of a computer to put it into RAM so it can be modified or printed. Sometimes things go wrong: we forget the file name; the computer file directory becomes corrupted so that the computer cannot find the file; or the file itself is corrupted. Similarly, learners can have difficulty retrieving information from long-term memory. The information may not be recalled at all, only partially recalled, or even misremembered as it is confused with other information. Common causes for such problems are infrequent recall/use of the information, and the passing of a long period of time.

It's Your Turn

How many of the ideas from these theories matched with those on your list?
Which of the theories would you like to learn more about?
Which do you see as being relevant to your own teaching situation?
Which do you think are important for your methods students to know about?
Which of these ideas about learning do you think would help explain Asmar and Peter's problem?

There are a lot of ideas in the learning theories we have just considered—far too many to incorporate directly into a science methods course, but many that still need to be considered in designing and implementing a course. In the next section we outline a view of learning that we, and many others, have found helpful in guiding instruction in both elementary science and in science methods courses.

A View of Learning

The following view of learning evolved over many years of working with in-service and preservice teachers, as I (Appleton) strove to find a way of making important constructivist ideas understandable to teachers when there was limited time.

Figure 1.1 portrays in an information processing flow diagram[1] the key features of the view of learning. Each of these features is then explained in more detail. The model portrayed by the diagram shows alternative pathways that may be taken by learners. The actual pathway followed by individual learners depends on their preferences about learning and their personal feelings at the time, their learned behaviors in formal education settings, and the influence of the social context in which the learning encounter occurs.

Overall Classroom Context

While the model portrayed in Figure 1.1 could apply to a range of learning contexts, it was developed from research in elementary science classrooms and also later applied to elementary science methods classes. The key elements of the model emphasize what is happening within the individual learner, but each of these is influenced and shaped by the social context.

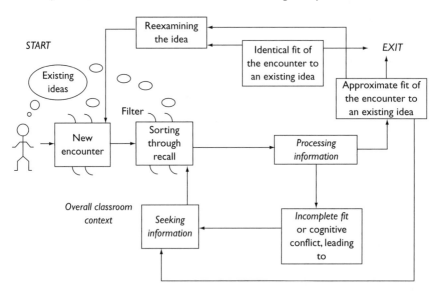

Figure 1.1 A View of Learning (source: Appleton, 1997).

1. Adapted with permission from Appleton, K. (1997). *Teaching science: Exploring the issues*. Rockhampton, Australia: Central Queensland University Press.

The classroom social context has considerable bearing on the individual learner's response, but other contexts such as the overall school social context, peer group contexts, and the home/family context all play a role. Influences of the social/cultural context are highlighted for each part of the model discussed.

Existing Ideas

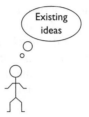

Learners coming to a learning situation bring with them their preexisting ideas, skills, and feelings that are organized in the mind as sets or clusters of ideas and experiences. As outlined earlier, these are called schemata and collectively constitute a cognitive structure. Schemata are developed through a combination of experience with the natural environment, language, and social interaction. They are shaped by the culture in which a learner lives. Specific schemata are developed for the school culture and classroom to which the learner belongs. Schemata dealing with related experiences are linked in the mind to form a complex web of interrelationships. Some of these preexisting ideas relating to the natural world may conform to the accepted views held by scientists, but some may be different. These alternative views are often called misconceptions or preconceptions. Schemata developed within school contexts can also be separate from everyday schemes, even though they may deal with similar natural events.

Preservice elementary teachers similarly bring with them views about schooling, learning, teaching, and science based on their own schooling, their life experiences, and formal science classes they have taken. Most would have had positive school experiences, because they have successfully negotiated their way through school to university. Those who came to university after finishing school have limited life experiences, but those starting later in life may bring with them extensive life experiences and less positive views about their schooling. The majority has negative views about science, based on their own schooling and formal science classes. Given the propensity for formal tests in schools, the prospective teachers' views of learning and teaching are likely to include notions of rote learning for success, and of testing as the main means of assessment.

New Encounter

The new encounter in the model is new in the sense that it is a further experience that occurs in a particular social and cultural context. In formal learning, the new encounter usually occurs within a school context, and is presented by the teacher as part of a pre-planned set of experiences and teaching strategies. Depending on the teaching strategy, the learner may be directly involved with the encounter individually or in small groups, or indirectly as part of the whole class. The learner attends to it because of social expectations within that context. A new encounter in elementary science might be a puzzling problem presented by the teacher, or preferably a demonstration or student activity that leads to new problems and activities that can be investigated. After the encounter, there would normally be several lessons where these problems and activities are investigated.

In a methods class, a new encounter would usually be planned for each session. It would normally focus on one or two key aspects of curriculum (such as a learning theory, a science concept, or a teaching approach). An advanced organizer (Ausubel, 1968) may be used to focus the prospective teachers' thinking such as, "Today we are looking at using discrepant events to initiate a unit of work in science." This would be followed by an actual discrepant science event demonstration and subsequent student investigations.

Sorting Through Recall

When some new encounter is experienced, existing ideas are used to make sense of it. The encounter will occur in a particular social context, which will trigger those memories, including feelings, recognized as relevant to that context. These will be the first to be called upon to make sense of the new encounter. The learner subconsciously searches through these schemata, starting with those deemed most relevant, for one that seems to fit the circumstances best. The learner takes cues from the context of the encounter, such as unit and lesson structures, and teacher comments, to aid in this search.

The preservice teacher encountering a discrepant event in the methods class would be searching memories to interpret what "discrepant event" might mean, and in viewing the demonstration would search through memories that featured similar objects to those used. Links to science ideas may also be sought.

Filter

The selected schemata act as a filter to incoming sensory data, determining which will be attended to and which will not be attended to (Osborne & Wittrock, 1983). That is, the learner may select to attend to particular aspects of the experience and ignore others. This selection may be by chance, by what is spectacular and attention getting, or by the learner's expectations of the lesson. Learners experiencing some emotional upset may miss most of the encounter because they are unable to attend to it fully. Learners in early puberty may be distracted by members of the other sex, and so attend more to them than to the encounter. This can occur in a science methods class too, where men or women compete for the attention of an attractive person of the opposite sex. The classroom context also influences the sensory input attended to and the memories activated to construct meaning for the experience. Consequently, everyday schemata may not be used. In sum, the social context and selected schemata together form a filter for the aspects of the encounter to which the learner attends, and for the interpretation of the encounter made by the learner.

An obvious filter in science methods classes relates to the scheduled time of day of the class: very early morning classes are affected by prospective teachers' unwillingness to arise early and the possible absence of breakfast; late afternoon classes are affected by other classes and social interactions during the day. Further, those who learnt to fear science during their schooling may have their engagement with the encounter affected by the filter of their fear. Those who view effective learning as rote learning for later regurgitation in a test may filter the encounter through this expectation because they are looking for the facts that they have to "learn."

Processing Information

> Processing
> information

The search for an appropriate explanatory memory continues until one that is relevant to the encounter is found. This search is a means of constructing meaning for the encounter, by processing the sensory information, selecting aspects considered important and relevant, and making sense of this using the selected recalled memories. Such processing may occur at a superficial level, termed Surface Processing by Biggs and Moore (1993). This involves accepting superficial similarities as an adequate match between memory and encounter, making little or no attempt to explore details to understand the encounter, and/or rote learning of aspects of the encounter required by the classroom context (such as a forthcoming test). It involves a minimum of mental processing of the available information into the learner's cognitive structure.

Biggs and Moore suggested that Deep Processing could alternatively occur, where learners actively try to reach understanding and make sense of the encounter by relating it to remembered schemata, using thought experiments, or generating analogies. To change or adjust ideas to fit with the new encounter, and therefore reach a deep level of understanding, often requires considerable effort. The Piagetian term accommodation is similar in meaning.

Which level of processing occurs depends on both the learner and the context. Learners who lack the cognitive skills to engage in deep processing must use surface processing. If the learner has the cognitive skills to engage in deep processing, Biggs and Moore (1993) suggested that learners will often use an Achievement Approach to learning (not indicated on the model), where they select what they perceive to be the most appropriate form of processing (i.e., surface or deep) to meet the demands of the learning context with the least effort. For example, if the expectations are that the lesson is important in a test involving mainly recall, then surface processing will probably be used. Processing will also be influenced by the social context, the teaching strategy, and the learner's emotional state. For instance, if the learner has experienced repeated failure in science, then it would not be considered worth the effort to engage in deep processing. Alternatively, if the teacher's questioning involves a rapid series of recall questions, the learner will effectively be prevented from engaging in deep processing.

Preservice teachers tend to be achievement oriented and pragmatic: they will only use the effort necessary for deep processing if they have to. If they can get away with surface processing, they will do so—particularly if this is what they consider successful learning involves. In our example of a discrepant event demonstration, several strands of sensory information would be processed simultaneously: the meaning of "discrepant event," the objects used in the demonstration, the outcome of the demonstration that does not align with memories and expectations (see degree of fit below), and the science ideas that may explain what happened (or did not happen).

Degree of Fit

| Identical fit of the encounter to an existing idea | Approximate fit of the encounter to an existing idea | Incomplete fit or cognitive conflict, leading to |

Once the information has been processed, there are three possible outcomes:

An Identical Fit

All noticed aspects of the new encounter are perceived by the learner as being an identical fit with an explanatory idea drawn from memory: from the learner's point of view, the encounter is fully explained by what is recalled. This in effect reinforces the prior learning and increases its status—highly desirable if indeed the recalled explanation is consistent with the scientific explanation. However, even though the learner perceives it as an identical fit, it is possible that the recalled idea fails to explain the encounter adequately from the teacher's point of view. This could be because the learner did not notice a particular aspect of the encounter, or because the learner overlooked some details of the recalled idea that did not fit with aspects of the encounter. Learners who feel that they have achieved an identical fit are most likely to exit from the learning situation and not attempt to engage further with the experience. Those who exit with a misconception reinforced face likely future learning difficulties.

Approximate Fit

Another problem can arise if the learner selects from memory an explanation that superficially explains the encounter, without checking for inconsistencies. That is, the new encounter may be perceived as being approximately explained by the remembered idea, even though the learner is aware that aspects are unclear and unexplained. This approximate fit is accepted as "good enough" and the learner exits from the learning experience with little if any learning gain, and potential future learning difficulties. The likelihood of this occurring is high in a competitive classroom, where the social climate encourages finishing a task quickly.

However, instead of exiting from the learning experience, some learners may keep an open mind by entertaining the idea as a possible, even likely explanation, and seek further information to clarify the situation (see below).

Incomplete Fit

The learner may recognize that no remembered ideas adequately explain the encounter, resulting in an incomplete fit and cognitive conflict (Piaget, 1978). A learner in a state of cognitive conflict experiences some degree of frustration or dissonance (Festinger, 1989), the motivational force that drives a learner to seek a solution to the learning situation posed by the new encounter. The level of cognitive conflict experienced by the learner depends on the nature of the encounter, the social context in which it occurs, the emotional state of the learner, and the memories available to potentially explain the encounter.

In our discrepant event example, the prospective teachers could have any of these three degrees of fit in linking to the meaning of "discrepant event": they may understand it because of previous reading or experience; they may have a fuzzy idea about what it means and accept that as "near enough"; or they may realize that they do not know what it means. Similarly, when viewing the demonstration, they may link to previous experience and science ideas that explain it, link to some vague science ideas that are accepted as an explanation, or be perplexed because they have linked to no satisfactory experience or idea.

Reexamining the Idea

> Reexamining
> the idea

A learner who has effectively exited from the learning experience because of a perceived identical fit or because a vague idea is considered adequate, may reengage with the learning situation if the classroom social context encourages this to happen. For instance, as learners cannot just leave the classroom, their enforced continuing presence during the lesson may result in their reexamining the idea. This could be triggered by further thought, reading about the topic, or a comment or action by a peer or the teacher. Careful planning of teaching strategies will enhance opportunities for learners to reengage with the learning experience instead of prematurely exiting from it.

Preservice teachers have had many years of schooling during which they have perfected the art of pretending to be engaged in the lesson, when they have really opted out and are, for example, waiting for the instructor to tell them the "answer" to rote learn.

Seeking Information

> Seeking
> information

The most common solution-seeking behavior arising from cognitive conflict is to seek further information. Any new information obtained essentially becomes a new encounter for a new cycle through the model. By obtaining further information, conflict is reduced as the learner processes the new information, modifies existing ideas, extends them, or constructs new ones (Osborne & Wittrock, 1983).

Possible ways of seeking information are (Appleton, 1993):

- exploring the materials vicariously, such as through a teacher demonstration;
- exploring the materials directly using hands-on;
- using the ideas of others who are external to the classroom, such as books, audiovisual and multimedia sources, and community experts;
- using ideas from the teacher;
- using ideas from peers, obtained one-to-one, in small groups, or in the whole class;
- waiting for an answer to be revealed, if the teacher maintains control over information flow and availability; and
- using unit, lesson, and teacher structuring cues, such as the topic from previous lessons, teacher actions, what the teacher says about the encounter, and what the teacher does not say.

The information sources used by a learner depend on the learner, the nature of the encounter, and the classroom social context—in particular, the teaching approach being used. Some learners may not have the information-accessing skills to use some sources effectively. For instance, the effective use of books depends on reading skills using a number of different reading styles, summarizing skills, and so on. Similarly, not all learners may recognize teacher lesson cues and structuring. However, the teaching strategies used by the teacher largely determine the information sources that a learner can access during a lesson. For instance, if the teacher uses a teacher demonstration, direct access via hands-on is not possible. Similarly, if the teaching strategy does not include small group work, the opportunities for obtaining information from peers are limited. The teaching strategy also controls what is done with the information obtained—whether it is confined to one person, a small group, or shared with the whole class.

The form of the information is also pertinent. Some information is not easily understood, let alone able to be related to the task at hand, because it is too complicated, or because it is hidden in a lot of extraneous information. Some learners therefore need help in identifying and accessing relevant information. Sometimes it may even be necessary to "translate" complex forms to simpler ones useful to the learner. A further aspect is the social climate of the classroom: information is most useful to a learner if there is a publicly agreed purpose for obtaining it, which is not at odds with the learner's private purpose. A grade 1 child's purpose may be to please the teacher, while an upper grade child facing a high-stakes test may wish to pass the test. A preferable, public purpose for information gathering is to use it to arrive at an explanation that is valid in terms of the evidence, and that is acceptable to the class. This means that information from all sources, including authority sources, must be considered contestable and testable. It requires specific action by the teacher to ensure that the appropriate social climate exists for such a view to be encouraged.

In our discrepant event example in a science methods class, those who do not understand the term "discrepant event" may simply ask the instructor for an explanation. When faced with the discrepant event demonstration, a common form of seeking information is to want to engage with the materials themselves and try to replicate the demonstration. For this to be orderly, and to provide an appropriate pedagogical model, the instructor may need to help them compare ideas, organize their thoughts, and plan different investigations into aspects of the discrepant event.

It's Your Turn

Do any aspects of this view of learning strike a chord with you?

Using the above outline as a basis, prepare an explanation that Noeline could use to explain to Asmar and Peter why their students have difficulty during examinations.

You may still be struggling with working through how all of this might apply to a methods course, so we offer some ideas in the next section.

An Example Application to a Science Methods Course

Keep in mind that this discussion about learning is intended to help clarify and perhaps challenge your own views about learning and teaching, a key component in the development of both science pedagogical content knowledge, and science methods PCK. So that you can see how some of the components of the above view of learning can influence science methods course design and delivery, we provide some further comments about how they were used to plan a course.

The existing ideas and feelings that prospective elementary teachers bring with them to science education classes were identified. A literature search revealed the low confidence levels of prospective elementary teachers in teaching science (e.g., Dooley & Lucas, 1981). Specific information about beginning teachers entering this course was also gathered (e.g., Appleton, 1991). For instance, it was discovered that almost all beginning teachers, prior to any instruction about science teaching, believed that activity-based or hands-on programs were most appropriate for the elementary school (Appleton, 1983). This information was a major consideration in setting the goals for the science methods course. Since each intake of prospective elementary teachers into the course was about 80 to 90% female, it was also considered that gender inclusive practices should have a high priority.

In devising appropriate new encounters for the science methods course sessions, several science content areas were selected, such as electric circuits. Other topics, such as learning theories and gender inclusiveness, were included as well. Topics were introduced by modeling them during class sessions, with specific reference to the modeling process to make the content explicit. For instance, gender inclusive strategies were used during class sessions, and the teachers were told how they were being used. The science content was presented through teaching approaches appropriate to elementary science teaching, which were also modeled during classes. In this way, the focus was on the teaching approaches, and the science content became an incidental but key component. It was assumed that this would reduce the already established anxiety levels many of the teachers felt about science. To highlight the potential for misconceptions being held or developed by school students, misconception research into the selected science topics was analyzed, and compared with the prospective teachers' own views in a non-threatening way.

To diagnose which memories students were using to interpret the encounters and other information, small group and whole group discussions were held. The discussions focused on both class-work—especially the modeling, and information the prospective teachers obtained from lectures, reading, and any other sources. The small group interactions provided opportunities for the instructor to listen to ideas the prospective teachers shared, and take remedial action as necessary.

Several facets to the prospective teachers seeking and processing information were considered. The process of a person explaining his/her ideas to a group required active recall of relevant memories, and forced each to try to assimilate and accommodate any new information. Since there was an interchange of ideas, the teachers were able to provide peer scaffolding as appropriate. Acting as a tutor in a scaffolding situation makes deep processing obligatory, and since the scaffolding is being provided by a peer, the recipient is more likely to attempt to engage with the content. The instructor was also able to engage in meaningful scaffolding exercises with small groups of the prospective teachers at the appropriate times, to assist them in deep processing of information. Another purpose of the group discussion was that it formed a major component of the gender inclusive approach for the unit, providing the women with a comfortable and preferred way of accessing knowledge and therefore reducing the reviving of feelings of failure and fear of science learned in school.

Further ideas about the application of learning theories that have been applied to methods courses are provided by Olson and Appleton (2006).

Conclusion

The theoretical views outlined in this chapter lay a key foundation for both the development of PCK for elementary science and elementary methods, and other important aspects of an elementary science methods course. Another foundation necessary for consideration is the orientation to teaching, or beliefs about teaching, that the instructor holds. We examine these in the next chapter.

> Noeline's coffee shop discussion with Peter and Asmar led to further discussions with her colleagues who were now keen to find out why their students made silly mistakes in assignments and exams, and what they might do to help their students learn more effectively. They began to share what they were picking up from their discussions with Noeline with other science faculty. This generated so much interest that Noeline was invited to present her ideas about learning at an informal presentation to the science faculty. Noeline reflected how a simple coffee shop conversation had resulted in her colleagues seeking to understand ideas about learning, with consequent changes in the way they taught.

It's Your Turn

Earlier in the chapter, we asked you to write a list of ideas about learning that you could think of. Review your list, and add to it new ideas that you have picked up from reading the further ideas discussed.

Have any of your ideas changed? In what way?

> Regarding Noeline being asked to present her ideas about learning at an informal presentation to the science faculty:
>
> *If you were Noeline, what would you say?*
>
> Prepare a presentation that you would use to help colleagues understand your views. Try to organize an opportunity to present it to some colleagues.

More to Explore

Abell, S., & Lederman, N. (2007). *Handbook of research on science education.* Mahwah, NJ: Lawrence Erlbaum.

Claxton, G. (1990). *Teaching to learn: A direction for education.* London: Cassell.

McInerney, D., & McInerney, V. (2006). *Educational psychology: Constructing learning* (4th ed.). Sydney, Australia: Pearson Educational.

National Research Council. (2005). *How students learn.* Washington, DC: National Academies Press.

Osborne, R., & Freyberg, P. (1985). *Learning in science: The implications of children's science.* Auckland, New Zealand: Heinemann.

Woolfolk, A. E. (2008). *Educational psychology* (11th ed.). New York: Merrill.

References

Appleton, K. (1983). Beginning student teachers' opinions about teaching primary science. *Research in Science Education, 13,* 111–119.

Appleton, K. (1991). Mature-age students—How are they different? *Research in Science Education,* 21, 1–9.

Appleton, K. (1993). *Students' learning in science lessons: Responses to discrepant events.* Unpublished PhD dissertation, Central Queensland University, Rockhampton, Australia.

Appleton, K. (1997). *Teaching science: Exploring the issues.* Rockhampton, Australia: Central Queensland University Press.

Ausubel, D. P. (1968). *Educational psychology: A cognitive view.* New York: Holt, Rinehart and Winston.

Bandura, A. (1986). *Social foundations of thought and action.* Englewood Cliffs, NJ: Prentice Hall.

Bandura, A. (1997). *Self-efficacy: The exercise of control.* New York: Freeman.

Biggs, J. B., & Moore, P. J. (1993). *The process of learning* (3rd ed.). Sydney, Australia: Prentice Hall.

Claxton, G. (1984). *Teaching and acquiring scientific knowledge.* London: Centre for Science and Mathematics Education, University of London.

Claxton, G. (1990). *Teaching to learn: A direction for education.* London: Cassell.

Dooley, J., & Lucas, K. (1981). Attitudes of student primary teachers towards science and science teaching. *Australian Science Teachers Journal, 27*(1), 77–80.

Festinger, L. (1989). A theory of cognitive dissonance. In S. Schachter & M. Gazzaniga (Eds.), *Extending psychological frontiers: Selected works of Leon Festinger.* New York: Russell Sage Foundation.

Flavell, J. H. (1985). *Cognitive development* (2nd ed.). Englewood Cliffs, NJ: Prentice Hall.

Furth, H. G. (1969). *Piaget and knowledge*. Englewood Cliffs, NJ: Prentice Hall.

Glasersfeld, E. V. (1987). *The construction of knowledge: Contributions to conceptual semantics*. Seaside, CA: Intersystems Publications.

Greeno, J. G. (1997). On claims that answer the wrong questions. *Educational Researcher, 26,* 5–17.

Hodgson, D., & Hodgson, J. (1998a). From constructivism to social constructivism: A Vygotskian perspective on teaching and learning science. *School Science Review, 79,* 33–41.

Hodgson, D., & Hodgson, J. (1998b). Science education as enculturation: Some implications for practice. *School Science Review, 80*(290), 17–23.

Kelly, G. A. (1955). *The psychology of personal constructs*. New York: Norton.

Kelly, G. A. (1963). *A theory of personality: The psychology of personal constructs*. New York: Norton.

McInerney, D., & McInerney, V. (2006). *Educational psychology: Constructing learning* (4th ed.). Sydney, Australia: Pearson Educational.

Nuthall, G. (1999). The way students learn: Acquiring knowledge from an integrated science and social studies unit. *Elementary School Journal, 99,* 303–341.

Nuthall, G. (2001). Understanding how classroom experience shapes students' minds. *Unterrichts Wissenschaft, 29,* 224–267.

Olson, J. K., & Appleton, K. (2006). Considering curriculum for elementary science methods courses. In K. Appleton (Ed.), *Elementary science teacher education: International perspectives on contemporary issues and practice* (pp. 127–151). Mahwah, NJ: Lawrence Erlbaum in association with the Association for Science Teacher Education.

Osborne, R., & Freyberg, P. (1985). *Learning in science: The implications of children's science*. Auckland, New Zealand: Heinemann.

Osborne, R., & Wittrock, M. (1983). Learning science: A generative process. *Science Education, 67,* 489–508.

Piaget, J. (1970). *Science of education and the psychology of the child* (D. Coltman, Trans.). London: Longman.

Piaget, J. (1974). *The child and reality: Problems of genetic psychology* (A. Rosin, Trans.). London: Frederick Miller.

Piaget, J. (1978). *The development of thought* (A. Rosin, Trans.). Oxford: Basil Blackwell.

Rogoff, B. (1990). *Apprenticeship in thinking: Cognitive development in social context*. New York: Oxford University Press.

Vuyk, R. (1981). *Overview and critique of Piaget's genetic epistemology 1965–1980: Vol. 1. Piaget's genetic epistemology*. London: Academic Press.

Vygotsky, L. S. (1962). *Thought and language* (E. Hanfmann & G. Vakar, Trans.). Cambridge, MA: MIT Press.

Vygotsky, L. S. (1978). *Mind in society: The development of higher psychological processes*. London: Harvard University Press.

Chapter 2

The Context for Elementary Science Teacher Preparation

> I met up with a colleague, Luke, at a conference and we decided to have lunch and chat about our semester. "I'm finding it really difficult to prepare my elementary education majors to teach science effectively," he began, "I'm the only science educator at my institution, and sometimes I feel like a one-man band! I wish that, like you, I had other people working with me to prepare these folks."
>
> "While it is nice having other science education colleagues at my institution, they're not the only ones who play a role in educating our elementary majors!" I responded.
>
> He turned to me, puzzled. "What do you mean?"

Who is Responsible for Educating Elementary Science Teachers?

While, as methods instructors, we may feel primarily responsible for the education of future elementary science teachers, in reality we are but a single piece of a complex system of teacher preparation. The effective functioning of this system relies on the collaboration among various stakeholders at both the program level and policy level. For example, during their university education, prospective teachers interact with scientists, science educators, other education faculty, and K-12 personnel (see Figure 2.1). The design, certification, and conduct of teacher education programs is influenced by various policies, created by members of these same groups, as well as professional and governmental organizations and policy makers. In essence, there are a variety of stakeholders in the education of prospective teachers, as well as a variety of contexts in which they are prepared. Though we have extracted the methods course out of context as the focus of this book, we acknowledge the task of the methods instructor in situating his/her course within the program and policy contexts. The purpose of

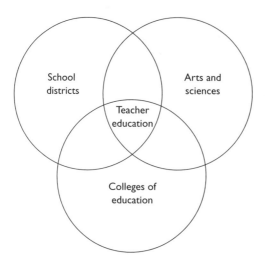

Figure 2.1 Collaboration of Stakeholders in Teacher Education.

this chapter is to encourage consideration, more broadly, of how the methods course fits into the bigger picture of elementary teacher preparation.

> ### It's Your Turn
>
> Make a list of the different groups of stakeholders in the preparation of *your* elementary science methods students. *For how many of these groups are you able to name specific individuals? With how many of these do you have regular communication? What opportunities for collaboration at your institution exist?*

The Policy Context: Goals and Standards for Elementary Science Teacher Education

In developing your science methods course, it is helpful to consider the broader landscape of policy that shapes and informs your role as an elementary teacher educator. Understanding where your course fits into this bigger picture, being sensitive to the external pressures you might face, and being conscious of what your students are doing beyond your own course to prepare for their future careers will be of benefit. There is no shortage of recommendations to be found in various reforms and policies regarding what teachers should know and be able to do—but such documents often provide little guidance in terms of practical suggestions for implementing these recommendations (Lavoie & Roth, 2001). The methods instructor must be aware of the relevant policies governing teacher preparation in his/her context, and design the methods course such that it supports teachers in meeting competencies laid out by the policy makers. In this section, we review some key policy documents relevant to elementary science teacher preparation, as well as the implications of those documents for the elementary science methods course.

One landmark science education policy document in the U.S. is the National Research Council's *National Science Education Standards* (*NSES*) (1996), which include standards related to science content, science teaching, assessment, professional development, programs, and systems. The "science teaching standards" outline what teachers are expected to know and be able to do. Each of the six broad standards includes specific definitions that can guide your thinking about appropriate goals for the science methods course. For example:

Teaching Standard B

Teachers of science guide and facilitate learning. In doing this, teachers:

- Focus and support inquiries while interacting with students.
- Orchestrate discourse among students about scientific ideas.
- Challenge students to accept and share responsibility for their own learning.
- Recognize and respond to student diversity and encourage all students to participate fully in science learning.
- Encourage and model the skills of scientific inquiry, as well as the curiosity, openness to new ideas and data, and skepticism that characterize science (NRC, 1996).

Many countries have policy bodies that guide the accreditation of our teacher education programs at the national or state level. In the U.S. for example, the Teacher Accreditation Council (TEAC) and the National Council for the Accreditation of Teacher Education (NCATE) are the main teacher education accrediting bodies. They regulate the components and content of teacher education programs, and universities that desire accreditation must address their policy standards. In some countries, accrediting bodies play a much more overt role, such as approving and regularly reviewing programs. As instructors, we need to know if there is content that we are expected to address in the science methods course as part of our unit's accreditation.

Some policy makers at the national, state/provincial, or local levels have defined standards for what beginning teachers should know and be able to do. For example, in the U.S., the Interstate New Teacher Assessment and Support Consortium (INTASC) Task Force on Teacher Licensing, established in the late 1980s, produced a common core of principles for teachers, defined by a set of knowledge, disposition, and performance standards. Later efforts included the development of subject specific standards for various content areas, including science (INTASC, 1992). These national efforts guided the development of teacher standards at the state level. For example, in the state of Missouri, the Missouri Standards for Teacher Education Programs (MO-STEP) (Department of Elementary and Secondary Education (DESE), 1999), although not subject-specific, define the knowledge and skills that beginning Missouri teachers are expected to possess.

Familiarity with the relevant policy documents, whether national, state, or local, can help you as a methods instructor figure out what to teach (see Chapter 5). For example, by identifying common standards in the policy documents relevant to your context, you can formulate a set of course goals and choose learning experiences that will enable your prospective teachers to meet the standards. In Table 2.1, we demonstrate how a goal for the elementary science methods course related to knowledge of assessment aligns with both the national and state standards.

It's Your Turn

Examine your syllabus—*to what extent is your course aligned with the relevant teaching standards in your context? What are the strengths and weaknesses in terms of alignment?*

Who accredits your institution's teacher preparation program? In what way are you accountable in terms of this accreditation process and your methods course?

The Program Context: Elementary Teacher Education

Elementary education majors typically take one science methods course. As such, the methods course is the primary vehicle through which prospective elementary teachers learn to teach science. Nonetheless, it is important to note that this experience is part of their broader college or university education. In this section, we consider how the elementary science methods course may be situated more broadly in the teacher preparation program at your institution.

The typical model for elementary teacher preparation in the U.S., Canada, Australia, and Europe is the 4-year certification program. However,

Table 2.1 An Example Alignment of Course Goals with Policy Documents

Elementary science methods course goal	U.S. National Science Education Standards (NRC, 1996)	Missouri Standards for Teacher Education Programs (DESE, 1999)
Demonstrate knowledge of assessment strategies and the ability to utilize assessment to guide science instruction at the elementary level.	Teachers of science engage in ongoing assessment of their teaching and of student learning. In doing this, teachers: • Use multiple methods and systematically gather data about student understanding and ability. • Analyze assessment data to guide teaching. • Guide students in self-assessment. • Use student data, observations of teaching, and interactions with colleagues to reflect on and improve teaching practice. • Use student data, observations of teaching, and interactions with colleagues to report student achievement and opportunities to learn to students, teachers, parents, policy makers, and the general public.	The pre-service teacher understands and uses formal and informal assessment strategies to evaluate and ensure the continuous intellectual, social, and physical development of the learner.

variations such as 2-year, 5-year, and post-baccalaureate certification routes also exist. Regardless of the length of program, prospective teachers will usually enroll in general education/liberal arts courses, science subject matter courses (discussed in the next section), and professional education courses (e.g., subject-specific pedagogy courses, like the science methods course, as well as foundational courses such as educational psychology) during their university education. Instruction of these courses is the responsibility of different groups of faculty, often in different departments or colleges within the institution or outside of it. For example, professional education coursework often involves collaboration with K-12 teachers in local schools.

Too often, teacher preparation programs are characterized by a lack of coherence and articulation across the general education, science education, and professional education curriculum strands (NRC, 1997). Additional criticisms include lack of deliberate recruitment and socialization experiences within the first 2 years of university education. As a consequence, the teacher preparation program, from the perception of prospective teachers, may simply consist of a checklist of unrelated courses to be completed on the way to earning a diploma and teaching certification. Given the institution-wide nature of teacher education, collaboration across departments and colleges is essential. For example, important issues such as facilitating the learning of learners with special needs or from diverse backgrounds may need to be reinforced throughout the teacher education program.

Other opportunities for collaboration may be built into the structure of the elementary education program. For example, professional education courses may be scheduled or arranged according to a "block" format in which elementary education majors, as a cohort, enroll in a set of subject-specific methods courses together. Though the elementary science methods course will most certainly build on ideas and concepts prospective teachers learn in their foundational courses such as educational psychology, the science methods course instructor may find him/herself more closely aligned with instructors of other methods courses (e.g., mathematics methods, reading methods) in terms of purpose and strategies. Collaborations with methods instructors from other subject areas in helping prospective teachers learn to teach may be facilitated through the use of a framework that helps them identify essential aspects of teaching that are common to the disciplines.

For example, one of us (Abell) once taught at an institution where methods courses were offered in a block. In this setting, each methods course instructor in the block shared a teaching and learning cycle that we made explicit to our students in our syllabi. The teaching cycle consisted of four components, linked through the process of reflection (see Figure 2.2). The assessment component of the framework relates to the teacher's task of gathering about student learning through a variety of sources. The evaluation component targets interpreting these data—looking for patterns in children's behaviors and thinking. The planning component leads to

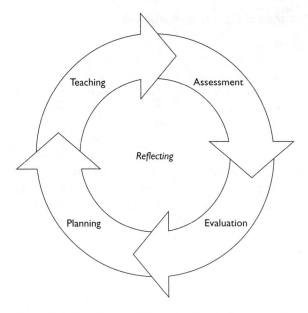

Figure 2.2 Teaching and Learning Cycle Framework.

planning instruction based on this evaluation to create opportunities to enhance, clarify, and build on children's knowledge. The final part of the framework, the teaching component, presents opportunities for the teacher to support children's movement toward more mature understandings. All four components are linked to a central feature of thinking like a teacher, reflecting. By sharing this framework across three different methods courses, we were able to help our students recognize and become comfortable with a process that they could apply across various subjects.

> ### It's Your Turn
>
> *What other courses do prospective elementary teachers take as part of their general education, science education, and professional education? What connections can you draw between these courses and your own?* Examine your course syllabus:
>
> *To what extent do you connect and situate your course within your students' broader teacher education program? How might you make this more explicit?*
> *Are there sets of core competencies or guiding principles for elementary teacher preparation that have been developed at your institution? How are these reflected in the syllabus and design of your course?*

The Program Context: Science Courses

Prospective teachers' knowledge of science is critical to effective science teaching (see Chapter 4). In this section we consider various models of instruction through which prospective teachers build their science content knowledge as part of their university education. University science courses provide prospective science teachers with their most recent opportunities to learn science as well as their most recent models of science teaching. As such, they can be powerful influences on how future teachers understand and appreciate science—in the ways they represent scientific knowledge and in the strategies they employ to help students acquire new scientific knowledge.

At your particular institution, prospective elementary teachers may be required to take one or more science courses for non-science majors. The status of undergraduate science education, particularly in introductory courses, has been heavily criticized in the U.S. (Sunal, Wright, & Day, 2004). In many cases, these courses fail to provide the kinds of learning experiences prospective teachers need to have a deep understanding of the content they will teach or appropriate models of pedagogy. Some science courses may be more desirable than others for prospective teachers to take, because of the ways they are designed and taught. Working closely with those who assist prospective teachers in making decisions regarding coursework is advisable.

While many faculty in the sciences assume that the responsibility for preparation of teachers resides solely in colleges of education, there is a growing recognition of the need for specialized science content courses for teachers (e.g., Beiswinger, Stepans, & McClurg, 1998; Duran, McArthur, & Van Hook, 2004; Edgcomb, Britner, McConnaughay, & Wolffe, 2008; Friedrichsen, 2001; Guziec & Lawson, 2004; McLoughlin & Dana, 1999) as well as the design of science curricula intended specifically for prospective teachers (e.g., McDermott, Shaffer, et al., 1996; Ukens, Hein, Johnson, & Layman, 2004). Such courses and curricula ideally should emphasize the content that teachers will be expected to teach, as well as model appropriate pedagogy, such as inquiry-based instruction. Additionally, they should focus on oft-neglected aspects of content such as the nature of science. "Teachers should be given the opportunity to examine the nature of subject matter, to understand not only what we know, but on what evidence and through what lines of reasoning we have come to this knowledge" (McDermott & DeWater, 2000, p. 245).

Specialized science courses for teachers may be planned with or without collaboration between science and education departments. These courses come in a variety of forms that may include, but are not limited to:

- science courses taught by education faculty (offered either in the college of education or the respective science department);

- science courses taught by science faculty in their respective science departments;
- science courses that are co-instructed by education and science faculty;
- science courses taught by faculty who are jointly appointed in science and education;
- integrated content and pedagogy courses for teachers.

For example, one of us (Hanuscin) experienced teaching specialized physical science courses for elementary education majors at two different institutions. While both courses were offered by the physics departments, one followed a traditional large-enrollment (200+ students) model with separate lecture and laboratory components, while the other followed an integrated lecture–laboratory model with an enrollment limited to 36 students. Currently, because she holds a joint appointment in both science education and physics, there have been enhanced opportunities for collaborations between the two program areas aimed at enhancing elementary science teacher preparation. Even if your institution does not have jointly appointed faculty, it is possible to establish open lines of communication and partnerships between faculty from different departments. It is likely that the faculty member who agreed to teach the course for teachers has a special interest in teaching, and thus would be open to such collaborations to enhance instruction.

By contrast, another of us (Appleton) was involved in a program that included science courses taught by education faculty, but which took an integrated content/pedagogy approach. The instructors held both science and education qualifications. This approach was adopted after extensive consultation with science faculty, school teachers, and prospective students; and reviews of student feedback on previous offerings, and reviews of research into the type of science content courses that prospective teachers find helpful.

It's Your Turn

What kinds of science coursework are your students required to take as part of their university general education requirements? What courses are available to them? When do they typically enroll in these classes—before, concurrently, or following the methods course?

Talk to advisors about recommended/required science courses for elementary education majors; make them aware of particular suggestions that you feel would enhance their preparation to teach science.

The Program Context: Field Experience and Partnerships with Local Schools

Additional stakeholders in the preparation of future teachers come from the K-12 educational system (see Figure 2.1). In Chapter 8, we discuss field experiences and practica in greater depth; however, we mention this now as one important program component for the methods instructor to consider. Researchers recognize a professional continuum of learning that spans preservice teacher education, induction of beginning teachers, and continued professional development (e.g., Feiman-Nemser, 2001). However, as Goodlad (1990) described, there is a general lack of collaboration and connectedness between schools of education and K-12 education. Universities typically regard preservice preparation as their task, with responsibility for new teacher induction resting with schools (Feinman-Nemser, 2001). We argue that effective elementary science teacher preparation must bridge this gap. Collaborations to design effective field placements and internships, as well as forms of simultaneous renewal that involve K-12 faculty in mentoring and instructional roles can benefit classroom teachers, university faculty, and prospective teachers.

Conclusion

At the beginning of the chapter, we posed the question, "Who is responsible for education elementary science teachers?" Clearly, the instructor of the elementary science methods course does not bear this burden alone. His or her work is carried out in the broader context of both policy and programmatic considerations. By identifying key stakeholders and establishing partnerships, methods instructors can benefit not only themselves, but also prospective teachers.

> I recognized my colleague, Luke, in the audience at our annual meeting's keynote address, and waited afterward to say hello. "Still feel like a one man band?" I asked.
>
> "Actually, talking with you last year really helped," he began. "I can't believe how much has changed in a year! Not only am I working more closely with other elementary education faculty to develop and focus on 'core competencies' for our students, but I made contact with a chemistry professor who enrolls a lot of education majors in her introductory course. We're in the process of submitting a grant to work on some improvements to the curriculum that I think will really benefit my students! If we're funded, we may be co-teaching a course in the future!"

> "It sounds like you're a lot happier now that you have collaborators," I commented.
>
> "Yes, I have to admit, teaching is often such a solitary endeavor that I've found the collegial conversations and reflection on my teaching as a source of renewal!"

More to Explore

The position statements of several professional organizations in the U.S. regarding science teacher preparation, as well as standards they have developed to inform course and program design are available online:

The Association for Science Teacher Education (ASTE) *Position Statement on Science Teacher Preparation and Career-long Development*. Retrieved September 17, 2009, from http://theaste.org/aboutus/AETSPosnStatemt1.htm.

The Association for Science Teacher Education (ASTE) *Professional Knowledge Standards for Science Teacher Educators*. Retrieved September 17, 2009, from http://www.lpi.usra.edu/education/score/ASTEstandards.pdf.

The National Science Teachers Association (NSTA) *Position Statement on Science Teacher Preparation*. Retrieved September 17, 2009, from http://www.nsta.org/about/positions/preparation.aspx.

In the U.S., accreditation of institutions that prepare teachers is conducted by two major organizations:

National Council for the Accreditation of Teacher Education (NCATE) (http://www.ncate.org/)

The traditional means of accrediting U.S. institutions that graduate teachers and present them for licensing. NCATE has produced Draft Accreditation Standards for Candidates in Elementary Teacher Programs, Standard 2b: Curriculum: Science.
Contact: NCATE, 210 Massachusetts Ave. NW, Suite 500, Washington, DC 20036-1023, 202-466-7496.

The Teacher Education Accreditation Council (TEAC) (http://www.teac.org/)

Founded in 1997, TEAC is a nonprofit organization dedicated to improving academic degree programs for professional educators, those who will teach and lead in schools, pre-K through grade 12. Membership represents a broad range of higher education institutions, from small liberal arts colleges to large research universities.

Contact: TEAC, One Dupont Circle NW, Suite 320, Washington, DC 20036, 202-466-7236.

References

Beiswenger, R., Stepans, J., & McClurg, P. (1998). Developing science courses for prospective elementary teachers. *Journal of College Science Teaching, 27*, 253–257.

Department of Elementary and Secondary Education (DESE). (1999). Missouri standards for teacher education programs (MoSTEP). Retrieved July 1, 2009, from: http://dese.mo.gov/divteachqual/teached/standards.htm.

Duran, L. B., McArthur, J., & Van Hook, S. (2004). Undergraduate students' perceptions of an inquiry-based physics course. *Journal of Science Teacher Education, 15*, 155–171.

Edgcomb, M., Britner, S. L., McConnaughay, K., & Wolffe, R. (2008). Science 101: An integrated, inquiry-oriented science course for education majors. *Journal of College Science Teaching, 38*(1), 22–27.

Feiman-Nemser, S. (2001). From preparation to practice: Designing a continuum to strengthen and sustain teaching. *Teachers College Record, 103*, 1013–1055.

Friedrichsen, P. M. (2001). Moving from hands-on to inquiry-based: A biology course for prospective elementary teachers. *American Biology Teacher, 63*, 562–568.

Goodlad, J. I. (1990). *Teachers for our nation's schools.* San Francisco, CA: Jossey-Bass.

Guziec, M. K., & Lawson, H. (2004). Science for elementary educators: Modeling science instruction for childhood education majors. *Journal of College Science Teaching, 33*, 36–40.

Interstate New Teacher Assessment and Support Consortium (INTASC). (1992). *Model standards for beginning teacher licensing, assessment and development: A resource for state dialogue.* Washington, DC: Council of Chief State School Officers. Retrieved July 1, 2009, from http://www.ccsso.org/content/pdfs/corestrd.pdf.

Lavoie, D. R., & Roth, W. M. (Eds.). (2001). Models of science teacher preparation: Theory into practice. Dordrecht, The Netherlands: Kluwer Academic Publishing.

McDermott, L. C., & DeWater, L. S. (2000). The need for special science courses for teachers: Two perspectives. In J. Minstrell & E. H. van Zee (Eds.), *Inquiring into inquiry learning and teaching in science* (pp. 241–257). Washington, DC: American Association for the Advancement of Science.

McDermott, L. C., Shaffer, P. S., & the Physics Education Group at the University of Washington. (1996). *Physics by inquiry* (Vols. I–II). New York: Wiley.

McLoughlin, A. S., & Dana, T. M. (1999). Making science relevant: The experiences of prospective elementary school teachers in an innovative science content course. *Journal of Science Teacher Education, 10*, 69–91.

National Research Council (NRC). (1996). *National science education standards.* Washington, DC: National Academy Press.

National Research Council (NRC) Committee on Undergraduate Science Education.

(1997). *Science teacher preparation in an era of standards-based reform.* Washington, DC: National Academies Press.

Sunal, D. W., Wright, E. L., & Day, J. B. (Eds.). (2004). *Reform in undergraduate science teaching for the 21st century.* Greenwich, CT: Information Age Publishing.

Ukens, L., Hein, W. W., Johnson, P. A., & Layman, J. W. (2004). Powerful ideas in physical science: A course model. *Journal of College Science Teaching, 33,* 38–41.

Chapter 3

Orientations to Teaching Science Teachers[1]

> "I have all of these ideas swimming in my head about teaching elementary science methods when I leave my doctoral program. It's hard to organize it all." My PhD advisee, Eun, was concerned about her future as a methods course instructor. "I've learned a lot from my apprenticeship, but I'm still not sure how to develop my own course." She had come to my office seeking advice from someone with the wisdom of practice developed over many years of teaching elementary science methods. Although I tried to make my teaching transparent and my reasoning explicit when she apprenticed with me, I felt unsure of how to guide her now. "Give me a day or two to think about this," I replied. Later in the day an idea came to me, so I dashed off an email to Eun: "I have an idea! Let's start with an exercise that will tell us more about your beliefs, purposes, and goals for teaching science teachers. See attached card sort task." Eun replied swiftly and positively. "I knew you'd come up with a plan to help me!" Within the hour, I had her responses to the card sort in front of me, and we were ready to interpret the findings.

Introduction

Science teacher educators hold beliefs, values, and assumptions about science teaching and learning, and about teaching science teachers, that help guide their planning and instruction. It is important for methods course instructors to uncover their ideas, and then take actions consistent with them. In other words, defining your orientation to teaching teachers can help you design and teach the methods course. In this chapter we discuss the construct of orientations to teaching teachers, provide several

---

1. Parts of this chapter were borrowed from Abell and Bryan, 1997.

examples of possible orientations to teaching science methods, and describe one orientation, which we call the reflection orientation, in greater detail.

What are Orientations to Teaching Teachers?

An orientation to teaching refers to a teacher's knowledge and beliefs about the purposes for teaching and serves as a conceptual map to guide instructional decisions (Magnusson, Krajcik, & Borko, 1999). Anderson and Smith (1987) introduced the idea of orientations to categorize different approaches to science teaching and learning (didactic, activity-driven, discovery, and conceptual change). An orientation represents a general view of teaching and learning science based on certain assumptions. A science teaching orientation can be defined (and differentiated from other orientations) according to teaching goals, characteristics of instruction, and views of science that are portrayed. Magnusson et al. expanded the Anderson and Smith list of orientations to teaching science from four to nine (adding process, academic rigor, project-based, inquiry, and guided inquiry orientations). The orientation construct suggests that how a teacher approaches science teaching is, in part, dependent on the teacher's overall philosophical stance toward science teaching and learning.

Orientation is a component of pedagogical content knowledge (PCK) (Abell, 2007; Grossman, 1990; Magnusson et al., 1999; Shulman, 1986) (see Introduction). We believe that orientations, like every other PCK component, take a parallel form as a type of knowledge for teaching teachers (Abell, Park Rogers, Hanuscin, Gagnon, & Lee, 2009; Abell, Smith, Schmidt, & Magnusson, 1996). While science teachers hold orientations to teaching science, science teacher educators hold orientations to teaching science teachers. We believe that an instructor's orientation to teaching science teachers serves as a filter through which their knowledge of science, science teaching and learning, and science teacher education is passed (Russell & Martin, 2007) as he/she designs and carries out instruction.

> ### It's Your Turn
>
> Follow the directions to Part I of the "Orientations to Teaching Science Teachers" card sort task (see "Tools" section). After doing the card sort, think about what your responses tell you about your beliefs about teaching science teachers. Read the next section and see which orientations your responses best fit.

Some Orientations to Teaching Science Teachers

Orientations can be illustrated, in part, by an instructor's purposes and goals for instruction. If we examine stated course purposes and goals, we derive insight into the instructor's beliefs and values. For example, our course descriptions states, "The purpose of this course is to equip you with the necessary knowledge, skills, and dispositions to be an effective teacher of science." This statement reflects the instructor's view that teachers need to have propositional and practical knowledge, as well as value that knowledge in order to put it into practice. Another of our course descriptions states course purpose in this way:

The purposes of this course include helping you to:

- clarify and refine your beliefs about teaching and learning science;
- become aware of children's ideas in science and how they influence learning;
- learn, practice, and reflect upon teaching strategies commensurate with your beliefs and knowledge about how children learn science;
- evaluate published elementary science textbooks and alternative curriculum materials;
- understand ways to assess student learning in science;
- gain sensitivity to the needs of diverse learners in science.

This purpose statement reveals that the instructor sees teacher knowledge as falling into several components—knowledge of learners, curriculum, instruction, and assessment—consistent with the PCK ideas presented in the Introduction. This purpose statement provides insight into what will be included and not included in the course.

In Chapter 5, we discussed some of the goals we have set for our courses. These goals also reflect our orientation to teaching teachers, including our beliefs about future teachers as learners. For example, in one of our course descriptions, one course goal is stated this way: "By the end of the semester, you should be able to present and defend your beliefs about elementary science teaching and learning." Underlying this goal is the belief that learners come to the science methods course with beliefs about science teaching and learning (see Chapter 4) that might be different from the beliefs in the policy documents guiding science education or from the science teacher educator. This instructor believes that one of the roles of the methods course is to help students revise and refine their orientations to teaching science in the elementary school. Another course description includes the following goal: "Apply appropriate professional knowledge to enhance teaching and learning." This broad goal implies that knowledge for teaching elementary science is a unique type of knowledge and that developing such knowledge is a major focus of the course and instructors.

Orientations to teaching science teachers can be exposed by the course goals and how the goals guide methods course organization and instruction. The card sort task (see "Tools" section) presents a number of scenarios from various methods courses. These scenarios reflect particular overarching goals for the methods course, methods of organization, and teaching strategies. They reveal possible orientations to the teaching elementary science methods course. Although individually a scenario might fit with several orientations, the card sort is designed so that when scenarios are combined, they represent one particular orientation. It is possible for an instructor to hold more than one orientation simultaneously. We briefly describe several orientations to teaching teachers, and then delve into one orientation in greater depth. There is not one right orientation to teaching teachers; however, we strive for consistency between our beliefs and our practices in science teacher education.

Topics Orientation

The goal of the science methods instructor who holds a topics orientation is to cover a number of important topics related to different aspects of science teaching—e.g., teaching science to diverse learners, using technology to teach science, teaching science through inquiry—during the elementary science methods course. Many methods courses are organized in terms of the content to be addressed, with each class session devoted to a particular science education topic. An instructor with a topics orientation to science teacher education often aims to cover each piece of the subject matter of science teaching.

Activity-Driven Orientation

The goal of the science methods instructor holding an activity-driven orientation is to help methods students experience science teaching in action from the perspective of a learner by modeling effective science activities. The methods students participate in one activity after another, assuming the role of elementary school students. The methods instructor aims for future teachers to develop their interest and confidence in science and to acquire a repertoire of classroom activities to use in their own teaching.

Pedagogy-Driven Orientation

The goal of the science methods instructor holding a pedagogy-driven orientation is to help methods students build a repertoire of science instructional strategies and approaches—e.g., cooperative learning, questioning, the learning cycle. The course is often organized by the instructional strategies being addressed. Preservice teachers read about the strategies and observe them being demonstrated by their instructor and/or by classroom teachers.

Teacher Inquiry Orientation

The goal of the science methods instructor who holds an inquiry orientation to teaching teachers is for future teachers to learn about science teaching and learning through carrying out a series of teacher-as-researcher tasks—e.g., finding out elementary children's science ideas, listening to a classroom discussion, carrying out an action research project. The instructor believes, through these teacher-as-researcher activities, that preservice teachers will learn how students learn science and make sense of best practices in science teaching.

> **It's Your Turn**
>
> Consider your responses to the "Orientations to Teaching Science Teachers" card sort task. *Which of the orientations discussed above, if any, fits your preferences?* Be aware that your ideas might fit into more than one orientation.

The Reflection Orientation and the Science Methods Course

The orientation that guides this book and our teaching of the elementary science methods course is the reflection orientation. The reflection orientation is grounded in the belief that learning to teach science, like learning science itself, is a process of reevaluating and reforming one's existing theories in light of perturbing evidence (Russell & Martin, 2007). Preservice teachers enter the elementary methods course holding ideas, beliefs, and values (Pajares, 1992) that form their personal theories about science teaching and learning. The goal of the science methods instructor who holds a reflection orientation is for students to confront and change their views of science teaching and learning through various opportunities for reflection. Instructors ask students to describe their ideas, beliefs, and values about science teaching and learning and offer experiences that help them clarify, confront, and possibly change their personal theories. Beginning teachers of elementary science need many opportunities to inquire into and think critically about science teaching and learning. A reflection orientation to science teacher preparation takes into account that future teachers learn about teaching science in a number of different contexts, each one of which can provide an opportunity for reflection and learning.

The reflection orientation can guide the design of an elementary science methods course. In the reflection-oriented course, preservice teachers engage in reflection within four unique but interrelated course contexts (see Figure 3.1). Throughout the methods course, students (1) reflect on

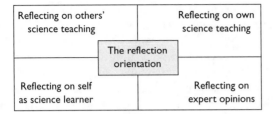

Figure 3.1 The Reflection Orientation.

others' science teaching via integrated media cases of conceptual change science teaching; (2) reflect on their own science teaching via field experiences in a partner school (see Chapter 8); (3) reflect on themselves as science learners via participation in science learning activities; and (4) reflect on expert opinions via course readings. Although we describe the course contexts as discrete components, each furnishing a singular opportunity for reflection on science teaching and learning, it is only by weaving all four components together that we create an integrated fabric for helping preservice elementary teachers become reflective teachers of science.

Reflecting on Others' Science Teaching

Cases create virtual worlds (Schön, 1987) within which preservice teachers can think about science teaching and learning. We developed the "Reflecting on Elementary Science" (Abell & Cennamo, 2004) (http://roes.missouri.edu/) videocases to serve as tools for reflection on science teaching and learning. The integrated media materials portray cases of teaching science for conceptual change (Russell & Martin, 2007; Scott, Asoko, & Leach, 2007). Three different cases show elementary teachers and their students as they progress through a series of lessons: fifth grade inclined planes, fifth grade levers, and first grade seeds and eggs. The cases illustrate a variety of naturally occurring classroom events including student investigations, small group interactions, student record keeping, large group discussions, and teacher demonstrations. A second audio track contains the teacher's reflections on each lesson. Other links support the videocase by providing text and graphics materials about the school, the teacher, the students, and the lesson plan, as well as background information about conceptual change science teaching, children's science ideas, and the scientific explanations involved in the lessons. We designed the materials to provide a great deal of flexibility for a variety of classroom formats. The materials can be used by a large group in a classroom setting to stimulate discussion and illustrate ideas. They can also be used by individuals or small groups independent of the instructor.

Infusing the integrated media cases into the elementary science methods course led to the development of a case-based pedagogy in which we used the videocases in conjunction with a series of reflection-about-teaching tasks (Abell, Bryan, & Anderson, 1998) to encourage systematic thinking about science teaching and learning. The reflection tasks ask students to clarify their entering personal theories about science teaching and learning, and compare their ideas to the instruction represented in the case. The case-based pedagogy was constructed to challenge students' personal theories and help them find intelligible and useful alternatives for teaching science. Thus, the videocase-based pedagogy helps elementary education majors clarify, confront, and revise their personal ideas, beliefs, and values about science teaching and learning. Many cases of elementary science teaching are available in print version and online (see Chapter 8).

Reflecting on One's Own Science Teaching

In a methods course designed with a reflection orientation, coupled with reflecting on others' teaching are opportunities for the methods students to reflect on their own science teaching by interacting with classroom teachers and their students in a field experience (see Chapter 8). For example, each section of the methods course might be partnered with an elementary school, where a group of teachers collaborates with the university on their own professional development in science teaching. Within a course section, each team of three methods students is matched with one elementary school teacher and her students for a variety of science teaching experiences. In another example, pairs of methods students go into the field during the last 3 weeks of the course and teach a unit of science instruction.

The field experience might begin with an introduction to the school and to the team's classroom. Methods teams could observe the elementary students in their classroom and on the playground and then compare their observations with those of teams from other grade levels. This provides methods students with a chance to discern the developmental characteristics of children, as well as to recognize developmentally appropriate science curriculum and instructional strategies. Concurrently, in the campus portion of the course, students reflect upon the first grade videocase and revise some of their ideas about what first graders can and should be expected to do in science. Thus the two contexts for thinking about science teaching and learning are linked.

Throughout the semester, several course assignments take place in the field. For example, students learn about a particular science curriculum by planning and teaching a science lesson with their host teacher, using the partner school's adopted science curriculum. Later in the semester, when studying children's ideas in science, methods students interview children in the partner class about a particular science concept. The methods team later applies this information about students to the development of a short learning cycle (Brown & Abell, 2007; Karplus & Thier, 1967). The field experience culminates with team teaching of the unit and corresponding assessment of student understanding. During this phase of the field experience, the classroom teacher serves as clinical supervisor, offering feedback and asking probing questions.

The methods students are also engaged in a series of written reflection tasks aimed at encouraging them to make sense of their teaching and helping them learn from the experience. Before the unit teaching we ask: What are your expectations for what will happen during your lessons? What do you envision happening with the students? Your teaching? What are your concerns/worries? After each lesson, we ask students to describe one classroom episode that was meaningful to them and tell why they chose that episode to describe. After the unit teaching has ended, we ask students to compare their initial expectations and concerns to the experience they had in the field, concentrating on ways their ideas about teaching and learning science have changed. The field experience contributes a personal and immediate teaching experience on which students of teaching can reflect. Students entering the elementary science methods course have a vision of themselves as science teachers that is in reaction to their experiences as science learners (Chapter 4). The field experience provides a new context in which to think about science teaching and learning. The field experience thus offers an opportunity to learn from experience. Enhanced by discussion and activities occurring in the other contexts of the course, the field experience becomes a powerful means for examining personal science teaching and learning theories.

Reflecting on Self as Science Learner

By doing and thinking and talking science within the methods course, preservice teachers inquire into their own understanding of science and their experiences as science learners, and use this inquiry to help make sense of their theories of science teaching and learning. For example, at the same time students are watching a videocase and becoming oriented to their field site, they are engaged in a long-term investigation of insects or of the moon (see Activities that Work 3 and 8). In their science investigations, they detect patterns and invent explanations about the phenomenon. They also reflect upon their own learning and the implications for science teaching. During class we often try to signal when the methods students are taking on the role of elementary science learner and when they are thinking like teachers. One way we cue the students is by using different parts of the room for different roles. For example, a circle of chairs at the end of the room signifies the scientists circle, where we discuss explanations and share evidence related to the science investigation at hand. When we move from the scientists circle to the methods lab tables, we tell the students it is time to move back to thinking like a teacher. Students generate many products during their science investigations—science notebooks, posters, diagrams, and charts. We try to keep these artifacts separate from, but connected to, the artifacts they develop from readings and discussions of science teaching.

Reflecting on Expert Opinions

Course readings and subsequent class discussions about science teaching and learning stimulate reflection about issues that methods students see played out in the other three contexts of reflection. When students are assigned a course reading, they are also asked to write about what they read in order to fit the reading into their ever-evolving system of ideas, beliefs, and values about science teaching and learning. They respond to the readings in the form of a "Reading Reaction" (Abell, 1992). The reading reaction consists of three sections—Before You Read, While You Read, and After You Read. Beginning questions activate students' prior knowledge and provide a purpose for the reading. The final question asks students to associate the reading with their prior knowledge: "Compare the ideas in this reading to related ideas you had before reading. How did your ideas change?" These questions can be completed and turned in online or done on paper.

The aim is for students to come to class having thought about what they read and how it relates to their personal theories as well as to other course experiences. In this way, science education authors are not held up as the sole authority for learning about science teaching and learning, but are regarded as one source of evidence for reevaluating and reforming personal theories. For example, while students are exploring issues of developmental appropriateness in science teaching in the field experience and Seeds and Eggs videocase, they read a chapter from Harlen (2000) titled, "Providing for Development 5–12." This reading explores developmental characteristics of different age groups and suggests age appropriate science activities. Methods students read and compare Harlen's ideas with what they observe in the videocase and field experience. They realize that general characteristics cannot be applied directly to individual children, and that the teacher has an important role to play in judging developmental appropriateness. They also read about scientists (e.g., Psycharis & Daflos, 2005) and compare what they read with the characteristics of children. This connects with the self as science learner course context wherein students engage in doing science and comparing themselves to scientists and to young science learners.

> **It's Your Turn**
>
> *What are the ways you ask students to be reflective practitioners in the elementary science methods course currently?* Think of one reflection context that you would like to develop more deeply in your course. Design one reflection assignment that will help your students learn from their experience as they synthesize course readings, discussion, and experiences.

Conclusion

The description of four components of one elementary science methods course illustrates how a course guided by a reflection orientation for teaching science teachers can be designed and implemented. Because we believe that learning to teach science involves clarifying, confronting, and expanding one's ideas, beliefs, and values about science teaching and learning, activities within the four reflection contexts require methods students to identify some of their ideas, beliefs, and values about science teaching and learning. The theories that they bring to the methods course are thus valued as knowledge; they begin to see their experience as having authority (Munby & Russell, 1994). Course activities have the potential to perturb some parts of students' theories and possibly help them accommodate new ideas, beliefs, and values about science teaching and learning. Furthermore, the instruction that they witness and take part in provides an alternative to the school science they have experienced.

In this way, the science methods instructor's orientation to teaching science teachers provides a road map of sorts for organizing the elementary science methods course. The orientation, grounded in a set of purposes and goals, will influence what gets emphasized and subsequently what gets learned. For example, if you hold an activity-driven orientation, you will organize your semester into various science activities in which your methods students engage. Students will learn those activities and potentially use them in their teaching. If you hold a teacher inquiry orientation, you will organize your semester into a series of teacher-as-researcher tasks from which you aim to help your methods students learn important principles of science teaching and learning. We believe that clarifying your orientation to teaching science teachers is an important process in becoming an instructor of elementary science methods courses, and in continuing to improve your teaching of science teachers over time.

> When I looked at Eun's response to the orientations card sort, I found that she had agreed, to some degree, with almost all of the cards. She needed to better define a framework for organizing her ideas. When we met again, I probed her thinking about science methods students as learners and her goals for teaching them. As we talked, she began to see nuances in the card sort scenarios and was able to eliminate some that did not fit her beliefs as well. As we remembered our experiences with the elementary methods course the previous semester, Eun saw connections between how we had organized and taught the course, and some of the card sort scenarios. She was able to take a step back, from focusing on specific strategies and assignments, to viewing the course as a whole. "I am beginning to see it now," she commented. "All of this really has to do with developing our students' knowledge for teaching science, and making sure they are reflective in every setting." I nodded. She had defined an orientation to teaching science teachers that would guide her in designing her own elementary science methods course.

Tools for Teaching Elementary Science Methods

Orientations to Teaching Teachers Card Sort Task[2]

Directions, Part I. Read each of the scenarios. As you read, think about your preferences for teaching the elementary science methods course. Sort the scenarios into three piles: Definitely agree with; Definitely disagree with; Not quite sure.

Scenarios for Teaching Elementary Science Methods

1. As a methods course instructor, you design an assignment for your methods students to read research about science teaching and learning. You ask the preservice teachers to participate as members of a research team to study some aspect of their field experience classroom.
2. As a methods course instructor, you engage methods students in elementary science activities and believe that they will glean science principles and pedagogical principles from participating.
3. For each class period, you prepare a PowerPoint and walk preservice teachers through the most important ideas on a particular topic.

---

2. The Card Sort Task is based on Friedrichsen and Dana (2003) and adapted from Musikul (2007).

4. As a methods course instructor, you encourage preservice teachers to examine and discuss the expectations for students in state or national standards. You ask teachers to research curriculum materials by measuring the extent to which curriculum materials reflect a specific group of content standards.
5. As a methods course instructor, you ask the methods students to watch a video of teachers teaching science and then to think, discuss, and write about their reactions to some of classroom events they witnessed.
6. As a methods course instructor, you ask the methods students to read their science methods course textbook about a particular instructional topic and come to class ready to discuss that topic.
7. As a methods course instructor, you demonstrate the effective science instructional strategy of cooperative learning to the class and ask them to add the strategy to their science teaching notebook.
8. As a methods course instructor, you engage methods students in an elementary science activity about electricity, and ask them to write up the activity description in their methods notebook so they will be able to reproduce it in their own classrooms.
9. As a methods course instructor, you ask the methods students to watch a video where first graders are studying seeds and eggs. Before watching, you ask them to predict how the unit will start. After watching the opening lesson, you ask students to compare their predictions to actual classroom events.
10. As a methods courses instructor, you organize your course calendar by a series of topics about different aspects of teaching science in the elementary school.
11. As a science methods course instructor, you develop a set of lessons for a third grade classroom about teaching concepts of sound. You use a group of these fun, and easy-to-do activities in the methods course.
12. As a methods course instructor, you ask the methods students to watch the classroom teacher in their field experience use questions during science class. Students come back and report on their observations of this science instructional strategy.
13. As a methods course instructor, you ask your preservice teachers to become researchers in the elementary science classroom, observing how students think about heat and temperature at various points in a science unit.
14. As a methods course instructor, you ask preservice teachers to keep a notebook of the best strategies for teaching elementary science they learn throughout the semester.
15. As a methods course instructor, you ask the preservice teachers to take part in an investigation of phases of the moon and regularly reflect on what they are learning about the moon and what they are learning about science teaching.

Directions, Part II. Look at the cards in your various piles. What do they have in common? Look at the list of orientations on the next page and the scenarios that fit each scenario. Which orientations appear most in your "Agree" pile? In your "Disagree" pile. Do you want to move any of your "Unsure" pile to one of the other piles? How would you characterize your orientation to teaching science teachers?

Orientations Scenario Cards by Orientation

Orientation	Related goals	Scenario cards
Topics	The goal of the science methods instructor is to cover a number of important topics related to different aspects of science teaching—e.g., teaching science to diverse learners, using technology to teach science, science teaching through inquiry—in the elementary science methods course.	3, 6, 10
Activity-driven	The goal of the science methods instructor is to help students experience science teaching in action from the perspective of a learner, by modeling effective science activities.	2, 8, 11
Pedagogy-driven	The goal of the science methods instructor is to help students build a repertoire of science instructional strategies and approaches—e.g., cooperative learning, questioning, the learning cycle.	7, 12, 14
Teacher Inquiry	The goal of the science methods instructor is for future teachers to learn about science teaching and learning through carrying out a series of teacher-as-researcher tasks—e.g., finding out elementary children's science ideas, listening to a classroom discussion, carrying out an action research project.	1, 4, 13
Reflection	The goal of the science methods instructor is for students to confront and change their views of science teaching and learning through various opportunities for reflection—about others' science teaching, about themselves as science learners, about themselves as science teachers, and about expert views of science teaching and learning.	5, 9, 15

References

Abell, S. K. (1992). Helping science methods students construct meaning from text. *Journal of Science Teacher Education, 3*, 11–15.

Abell, S. K. (2007). Research on science teacher knowledge. In S. K. Abell & N. G. Lederman (Eds.), *Handbook of research on science education* (pp. 1105–1149). Mahwah, NJ: Lawrence Erlbaum.

Abell, S. K., & Bryan, L. S. (1997). Reconceptualizing the elementary science methods course using a reflection orientation. *Journal of Science Teacher Education, 8*, 153–166.

Abell, S. K., Bryan, L. A., & Anderson, M. A. (1998). Investigating preservice elementary science teacher reflective thinking using integrated media case-based instruction in elementary science teacher preparation. *Science Education, 82*, 491–510.

Abell, S. K., & Cennamo, K. S. (2004). Videocases in elementary science teacher preparation. In J. Brophy (Ed.), *Using video in teacher education: Vol. 10. Advances in research on teaching* (pp. 103–129). New York: Elsevier Science.

Abell, S. K., Park Rogers, M. A., Hanuscin, D., Gagnon, M. J., & Lee, M. H. (2009). Preparing the next generation of science teacher educators: A model for developing PCK for teaching science teachers. *Journal of Science Teacher Education, 20*, 77–93.

Abell, S. K., Smith, D. C., Schmidt, J. A., & Magnusson, S. J. (1996, April). *Building a pedagogical content knowledge base for elementary science teacher education*. Paper presented at the National Association for Research in Science Teaching, St. Louis, MO.

Anderson, C. W., & Smith, E. L. (1987). Teaching science. In V. Richardson-Koehler (Ed.), *Educators' handbook: A research perspective* (pp. 84–111). New York: Longman.

Brown, P. L., & Abell, S. K. (2007). Examining the learning cycle. *Science and Children, 44*(5), 58–59.

Friedrichsen, P., & Dana, T. (2003). Using a card sorting task to elicit and clarify science teaching orientations. *Journal of Science Teacher Education, 14*, 291–301.

Grossman, P. L. (1990). *The making of a teacher: Teacher knowledge and teacher education*. New York: Teachers College Press.

Harlen, W. (2000). *Teaching, learning, and assessing science 5–12*. London: Paul Chapman Publishing.

Karplus, R., & Thier, H. D. (1967). *A new look at elementary school science*. Chicago: Rand McNally.

Magnusson, S., Krajcik, J., & Borko, H. (1999). Nature, sources and development of pedagogical content knowledge for science teaching. In J. Gess-Newsome & N. G. Lederman (Eds.), *Examining pedagogical content knowledge: The construct and its implications for science education* (pp. 95–132). Boston: Kluwer.

Munby, H., & Russell, T. (1994). The authority of experience in learning to teach: Messages from a physics methods class. *Journal of Teacher Education, 45*, 86–95.

Musikul, K. (2007). *Professional development for primary science teaching in Thailand: Knowledge, orientations, and practices of professional developers and professional development participants*. Unpublished PhD dissertation, University of Missouri, Columbia, MO.

Pajares, M. F. (1992). Teachers' beliefs and educational research: Cleaning up a messy construct. *Review of Educational Research, 62*, 307–332.

Psycharis, S., & Daflos, A. (2005). Webbing through science history. *Science and Children, 43*(2), 37–39.

Russell, T., & Martin, A. K. (2007). Learning to teach science. In S. K. Abell & N. G. Lederman (Eds.), *Handbook of research on science education* (pp. 1151–1178). Mahwah, NJ: Lawrence Erlbaum.

Schön, D. (1987). *Educating the reflective practitioner: Toward a new design for teaching and learning in the professions*. San Francisco: Jossey-Bass.

Scott, P., Asoko, H., & Leach, J. (2007). Student conceptions and conceptual learning in science. In S. K. Abell & N. G. Lederman (Eds.), *Handbook of research on science education* (pp. 31–56). Mahwah, NJ: Lawrence Erlbaum.

Shulman, L. S. (1986). Those who understand: Knowledge growth in teaching. *Educational Researcher, 15*(2), 4–14.

Chapter 4

Understanding the Elementary Science Methods Student

"I'm really disturbed about several comments on my course evaluations from my elementary science methods course last semester," Jan, a new colleague, confided. "A couple of students wrote that they didn't really learn how to teach science—I was caught off guard! What am I doing wrong?"

"I've encountered this same problem before," I empathized. "I think it has a lot to do with the expectations students have for a 'methods' course—it seems as if some of my students want me to just tell them step-by-step how to teach ... One of the things I do in the course is discuss explicitly how one learns to teach—and I pose the rhetorical question—'If learning to teach science were simply following a series of steps, wouldn't *every* teacher be an excellent science teacher?' This seems to shake them up as they recall their past science experiences, and ineffective science teachers they had. Getting them to question their ideas is an important first step, but first you have to know what their ideas are."

"What are good ways to find out what their initial ideas are?" she asked.

Introduction

Developing your own PCK necessarily involves developing an understanding of learners—in this case, the prospective elementary teachers you will be teaching in your methods course. In the vignette above, we see the consequences of overlooking students' prior knowledge and expectations. It's easy for teachers to turn their focus toward their own teaching and what they are doing—as a new teacher educator, I was once chided by a colleague: "If you'd stop worrying about what *you're* doing and start focusing on what your *students* are doing, you'd be fine!" The notion underlying his advice was that knowing my students should guide my instructional decisions.

Though elementary education students are unique individuals, they nonetheless share aspects of their background and experiences that contribute to their views of science and science teaching. Many researchers have identified and explored the importance of preservice elementary teachers' knowledge, beliefs, and attitudes about science, science teaching, and science learning. In this chapter, we'll describe some of these factors to help you develop an understanding of the "typical" elementary science methods student. In our discussion, we will identify common points of resistance to learning to teach science that you may encounter as an instructor of an elementary science methods course. We acknowledge, however, that you will teach students with diverse interests and abilities, and that you cannot develop your knowledge of learners and PCK for teaching elementary science teachers by simply reading this chapter; therefore, we will also suggest strategies for you to use in your methods course to further develop knowledge of your students and their particular ideas about science and science teaching. In turn these will help you select appropriate curriculum, instruction, and assessment strategies to support student learning (see Chapters 5, 6, and 7).

Prospective Teachers as Learners

As discussed in Chapter 1, learners bring with them preexisting ideas, skills, and feelings that are organized in the mind as sets or clusters of ideas and experiences, or schema. Your methods students are learners not only of science, but also learners of science teaching. Preservice teachers' own beliefs, knowledge, and attitudes toward learning science have important implications for the development of their knowledge, attitudes, and skills for teaching science. A positive motivation and attitude toward learning science may translate to increased interest in teaching science, which will be to the ultimate benefit of elementary students, as the attitudes of teachers toward science play a vital role in shaping their students' motivations and attitudes toward science. Furthermore, low outcome-expectancy (a teachers' belief in students' ability to learn science) is related to teachers' own perceived deficiencies in background knowledge and lack of success in learning science (Ramey-Gassert, Shroyer, & Staver, 1996).

As learners, prospective teachers' motivations may either be directed toward performance goals (doing well on the assignment) or learning goals (improving their knowledge and skills). Those with the former may be preoccupied with grades and afraid of taking risks, while those with the latter may tackle challenging tasks more readily. You may find students in your methods course who primarily have a performance motivation to be resistant to participate in activities in which the right answer is not immediately apparent. They may be more focused on figuring out what your expectations are or what answer you're seeking versus figuring out the problem or task.

Prospective teachers' own experiences learning science serve as a powerful influence on their developing understanding of science teaching. As successful graduates of that system, they often have expectations to teach science in ways similar to the way they were taught—using textbooks and highly structured investigations that were laid out for students step-by-step. They may enter the science methods course expecting to learn a "script" to follow, and as a result they may be resistant to open-ended and inquiry methods and reform-based practices. Furthermore, your methods students may be operating in their self-acknowledged weakest area, having never found themselves successful as learners of science. Rather than viewing prospective elementary teachers from a deficit perspective, however, we wish to emphasize the beliefs, prior knowledge, and expectations held by your students as starting points for meaningful instruction in your methods course. Challenging prospective teachers' beliefs, knowledge, and expectations can be a powerful way of encouraging them to restructure their ideas about what it means to learn and to teach science (Bryan & Abell, 1999), and can serve as a beginning point for reshaping their practice (Gunstone, Slattery, Baird, & Northfield, 1993). In the sections that follow, we look more closely at the elementary methods students in terms of their attitudes and PCK for science teaching. We'll examine the research base regarding the attitudes toward science, orientations toward science teaching, and the various knowledge bases that prospective elementary teachers bring with them to the methods course.

> ### It's Your Turn
>
> Before you read the sections that follow, take a moment to reflect on your own ideas about a "typical" science methods student. *How would you characterize this typical student's attitudes, knowledge, and skills related to teaching science? What factors informed your knowledge of learners up to this point?*

Prospective Teachers' Attitudes toward Science and Science Teaching

Elementary teachers' anxiety and negative attitudes about teaching science have been well documented. For example, in a recent survey, 77% of U.S. elementary teachers considered themselves "well-qualified" to teach language arts/reading, while fewer than 3 in 10 indicated that they felt "well-qualified" to teach science (Weiss, Banilower, McMahon, & Smith, 2001). It would be natural to assume that the students in your methods course are similarly anxious about science and teaching science. While that may be true to some degree, research over the past two decades has challenged this assumption. Some studies have suggested that elementary education students with negative attitudes toward science may actually be a minority (Jarrett, 1999; Young & Kellogg, 1993), and most methods students will initially have a moderate interest in teaching science (Appleton, 1995) despite their anxiety. You may find that though your students have low levels of science knowledge and less confidence in their ability to teach science than other subjects, they are nonetheless highly motivated to improve their knowledge for teaching science. However, such motivation does not necessarily translate into open-mindedness; your students come to you already possessing a set of beliefs about science, and with a particular orientation toward teaching science. The more firmly held these beliefs are, the more difficult it can be for students to consider new perspectives.

> ### It's Your Turn
>
> Ask your students to rate their attitude toward science on a scale of 1–10; 10 being best.
>
> *Do you have distinct groups in terms of those who like/dislike science or feel ambivalent about it?*
> *How do your students rate their attitude toward teaching science? Other content areas?*
> *How might knowing this information help you group students to work together during projects and activities? How might they benefit from exposure to a different perspective?*

Prospective Teachers' Orientations toward Teaching Science

As discussed in Chapter 3, orientations toward teaching science refer to a teacher's knowledge and beliefs about the purposes and goals of science education for a specific target group. These orientations play a role in guiding teachers' decision making, and therefore to a large extent will influence classroom norms, instructional strategies, and methods of evaluation (Magnusson, Krajcik, & Borko, 1999). For prospective teachers, incoming orientations similarly serve as a filter—allowing them to embrace some material presented in the methods course, while contributing to their reluctance/resistance to embrace others.

It cannot be overstated that students are not blank slates; research has repeatedly demonstrated that students enter science teacher preparation having already formed specific ideas about teaching and learning science. These orientations have been developed over many years of experiencing science teaching as learners, from primary school through higher education. As such, it can be difficult to change prospective teachers' ideas about teaching and learning, and thus difficult to affect their practice. Many view the way they were taught science as a model for their own instruction, particularly if they were successful students. Of course, your students may define success in terms of the grades they received, whether or not these grades were an indicator of meaningful learning. This can be problematic in cases where the type of science instruction your students experienced in elementary school was primarily didactic and based on rote learning of scientific information. Such experiences may contribute to the belief that teaching science is a fairly straightforward process of following a script. As a result, prospective teachers may come to the methods course with the expectation that they will be provided with this script, and step-by-step "methods" for teaching science to elementary learners. They may also expect their students to complete activities by following a carefully outlined step-by-step procedure—portraying science as simply a matter of following directions—a common misconception about the nature of science. On the other hand, many of our students have come to the methods course dissatisfied with their own science learning experiences. This opens the door to exploring other ways of teaching and learning science.

> **It's Your Turn**
>
> *Have your methods students reflect on their own experiences as a learner—do they consider themselves successful in science? On what basis?* Discuss your expectations with students, and develop a common vision of what it means to be successful in your methods course.

Prospective Teachers' Knowledge for Science Teaching

The goals of the methods course (see Chapter 5) center on the notion of developing prospective teachers' subject matter knowledge and PCK for teaching science. This necessarily involves developing a diversity of knowledge bases for teaching. In the sections that follow, we summarize the implications of research related to the various knowledge bases that comprise prospective teachers developing PCK for teaching science.

Subject Matter Knowledge: The Nature of Science

Understanding of the nature of science, or what characterizes science as a particular way of thinking, is essential to scientific literacy. Unfortunately, many elementary education students think of science as a course they took in school or the body of knowledge found between the covers of a science textbook—they rarely have an appreciation of the scientific enterprise more broadly. Their ideas about the nature of science, shaped by their school science experiences, are often grossly distorted. Research throughout past decades has confirmed the notion that prospective teachers lack a firm understanding of the nature of science (for a summary, see Lederman, 2007). For many, science may be seen as the way to understand the world, to the exclusion of other ways of knowing. This can create conflict between students who ascribe to a religious worldview, and believe that science poses a threat to their beliefs.

To some of your students, scientific ideas may be viewed as irrevocable truth, based solely on observations of the natural world, to the exclusion of more subjective human elements, such as human creativity and imagination, beliefs, and opinions. Indeed, scientists may be viewed as being particularly unbiased and objective. This further serves to dehumanize and depersonalize science—it would not be uncommon for us to encounter a student who says they shied away from science as a career choice because they considered themselves more a "people person."

Inductivist views of science, in which accumulation of evidence through the scientific method constitutes proof or truth, prevail. The caution here is that individuals with these views may be more prone to accept without question claims that are "scientific" or made by scientific authorities. Rather than taking a skeptical stance toward scientific claims, students may instead rely on others to tell them what is right or correct. This is especially problematic when coupled with a lack of science content knowledge.

> **It's Your Turn**
>
> *What image of science do you want your methods students to develop in your course? What experiences will you provide them to help them overcome their stereotypes and misconceptions?*

Subject Matter Knowledge: Science Content

Prospective teachers, even those with a more extensive science background, have serious weaknesses in their understanding of science concepts (e.g., misconceptions about science topics that in many cases are very similar to those of younger students; Abell, 2007). Although the relationship between science teachers' content knowledge and their teaching practice is contested, there is evidence in the research literature that elementary teachers with a weak content background tend to avoid teaching science (Tilgner, 1990). Unlike secondary science education majors, elementary education majors are prepared as generalists, and typically take no more than the general education requirements in terms of science coursework (often three courses or nine semester hours in the U.S.). While one obvious suggestion is to increase the number of science courses elementary majors take, research illustrates that this will not necessarily have the intended effect. Moore and Watson (1999) found the majority of elementary education majors were not positively influenced by their college science experiences. Schoon and Boone (1998) found, however, that science content courses that were specifically designed for elementary education majors were more likely to be of value than those intended for non-majors. If your college or university has such courses, collaboration with the faculty who teach them can be mutually beneficial (see Chapter 2). For example, if the science coursework is prerequisite to the methods course, learning experiences from science courses can be a shared experience for your students, and as such can serve as a jumping off point for teaching a particular pedagogical concept (Hubbard & Abell, 2005). Similarly, science instructors may be able to emphasize common misconceptions in relation to various concepts they teach.

Even with additional preparation, the reality is that elementary education majors will likely enter their teaching careers without a good understanding of all the science content they are expected to teach. This causes a dilemma for the methods instructor, given his/her course is intended to focus on pedagogy, rather than science content. Nonetheless, there are opportunities to emphasize science content within learning experiences intended to focus on specific pedagogical principles (see Part II). We highlight several considerations below.

Knowing that your methods students may not have a firm grasp of concepts in the elementary curriculum, you may find it beneficial to engage them in science learning experiences. In addition to modeling particular pedagogical strategies, these lessons can address common misconceptions your students have about the content. Though you may want to use such experiences to model instruction appropriate for elementary learners, it is nonetheless important that you engage your methods students at a level appropriate for them—as adult learners. One strategy for doing this is to have students alternate "thinking like a learner" and "thinking like a teacher"—making it clear when you want them to assume different roles in the activity. By doing so, you can help students make the difficult transition from learner to teacher.

> **It's Your Turn**
>
> *What is one science concept you expect to teach your prospective teachers in your methods course? What should your students already know in order to learn this concept? What misunderstandings are they likely to have? What do you expect your students to find difficult about learning this concept?*

Knowledge of Instructional Strategies

Methods students may be unfamiliar with instructional strategies and approaches such as inquiry, conceptual change teaching, and the learning cycle. Indeed, their unfamiliarity with such methods might lead them to falsely conclude inquiry to be a "new" approach, despite its origins in the early 1900s! Prospective teachers may not easily adapt to new instructional strategies and approaches, either due to their own unfamiliarity and lack of models of these types of instruction, or a perception that they are not well supported in schools. The latter may be reinforced by their field experiences. Methods students' lack of knowledge of instructional strategies may become apparent when they are asked to create lesson plans. However, their knowledge can be built through implementing various Activities that Work in the science methods course (see 2).

> **It's Your Turn**
>
> *Where do your students locate resources to plan instruction? How do they evaluate their quality?*
>
> Ask your methods students to bring in several examples of science lessons they feel are appropriate for elementary learners. In small groups, have students share the criteria by which they chose their resources, as well as where they located resources.
>
> *How can you use students' ideas as a starting point for instruction?*

Knowledge of Curriculum

Just as prospective elementary teachers may enter their methods course expecting to learn a "script" for how to teach science, they may also expect to be provided a curricular "script" once they enter the classroom. Southerland and Gess-Newsome (1999) found elementary student teachers took their science curriculum as given, without questioning the topics to be taught. In the methods course, it is important not only to familiarize students with common curricular materials used in the elementary classroom (see Chapter 5) but also to help them develop a broader understanding of the goals and objectives for elementary science, so that they can make informed decisions about how to use curricular materials to meet those goals. Deciding which concepts are critical to understand, and how deeply particular concepts need to be covered, are likely to be a challenge to novice teachers.

Not having a large pedagogical repertoire, students may be likely to rely on Internet searches to locate curriculum resources and lesson ideas. Some methods students may be less discerning than others in choosing which resources to use. Without a good understanding of science or science teaching, prospective teachers may unknowingly rely on materials with inaccuracies in content or instruction. For example, prospective teachers may be likely to focus on superficial features of the activity, such as whether it would be fun or easy to do—without consideration of whether the activity led to a deep conceptual understanding in their students. It is important for the methods instructor to help students critically evaluate curriculum resources for use in their teaching.

It's Your Turn

Think about your own experiences. *As a teacher, how do you know what topics are appropriate for your students? What sources of information inform your expectations?*

Now consider your methods students—*how will they formulate expectations for their elementary learners?*

Invite students to generate a list of topics or concepts (or provide a checklist from which they can select) they think would be appropriate to teach in an elementary classroom. Have them compare their ideas to various national, state, and/or local standards and benchmarks.

Knowledge of Learners

In our experience, we find prospective teachers often have low expectations for the learning of their future students. As a result of their own lack of understanding of science content, they may be likely to view certain concepts as "too difficult" for elementary-age learners, or they may believe that fostering positive attitudes toward science is more important than promoting deep conceptual understanding at this level. Their knowledge of elementary learners may be based on fuzzy memories of their own elementary science experiences—which are more often than not impoverished. In comparison to expert teachers, prospective teachers are more likely to be unaware of alternative conceptions (or "misconceptions") students are likely to hold about a topic, as well as specific difficulties they may have with learning particular content (e.g., Akerson, Flick, & Lederman, 2000). Therefore, as part of their preparation to teach, methods students should become familiar with common misconceptions held by elementary learners as well as tools and strategies for assessing students' ideas (see "More to Explore"). Just as we emphasize getting to know *your* methods students", it is important to assist your methods students in getting to know elementary students—their ideas, interests, abilities, and particular ways of thinking about science. Helping your methods students develop appropriate expectations can be a challenge, particularly if they are not seeing high-quality science instruction taking place in the schools they visit in their field experiences (see Chapter 8).

Knowledge of Assessment

Though there is a general lack of studies on future and beginning teachers' knowledge of assessment, in our experience we find this to be one of the weakest areas of knowledge for our students. They may lack knowledge of what knowledge and skills are important to assess, and focus instead on criteria such as neatness, completion, or effort to assign grades to student work products. Many will have limited knowledge of assessment strategies, and associate assessment with tests and quizzes, viewing assessment as solely summative in nature. An important function of the methods course is to equip them with assessment knowledge and skills; however, students should also grasp the multitude of purposes and uses of assessment, the ethical dimensions related to each, and their roles as assessors. We also need to help them think more deeply about how teachers learn from assessment data and how those data can inform their instruction.

Understanding Your Methods Students

Getting to know your students is especially important at the beginning of the course, in terms of knowing where to begin with your students and how to proceed to where you would like your students to be by the end of the course. Activities designed to elicit students' prior ideas can also facilitate students' learning, by making them aware of their own thinking and exposing faulty reasoning, biases, or misconceptions. They can serve an additional purpose of helping build community and develop a shared vision with your students. Though we have provided information about what you may be able to expect in terms of your methods students' incoming ideas, this is no substitute for understanding your particular students. In Chapter 7 we discuss assessment in the methods course in depth. In this chapter we highlight several examples of ways you can assess your students' incoming ideas about science and science teaching, in order to develop your understanding of learners. Each method is intended to help you identify points of resistance to learning to teach, as well as offer strategies for addressing those.

Draw-A-Scientist

The Draw-A-Scientist Test (Chambers, 1983) is a popular assessment activity we have used with success in our methods courses. On the first day of classes, students are asked to draw a picture of a scientist doing science. Much like younger students, prospective elementary teachers often draw stereotypical images—overwhelmingly white males in lab coats, working alone and indoors. A gallery walk, in which drawings are posted for students to examine for patterns, allows students to notice similarities and differences among their perceptions of science and scientists. Those with relatively positive attitudes toward science benefit from the realization that others may feel negatively about science and vice versa. We use classroom discussions to consider the sources of students' impressions and reasons for their present attitudes toward science. Often, students will raise the question as to whether their drawings are a "fair" portrayal—and whether this is the image of science they want to pass along to their students. Thus, this activity can also serve as a useful tool for helping to shape students' goals as science teachers. For students who have little firsthand experiences with science and scientists, it can be important to provide them with experiences that can challenge their stereotypical images of science. Luckily, a college or university community often has real, live scientists whom students can interview to find out more about what scientists actually do! We have found our colleagues more than willing to take the time out of their day to visit with a future teacher in the interest of promoting public understanding of science. Revisiting students' drawings after their visit can help students further question the validity of their perceptions, and recognize where these may be based on limited information.

It's Your Turn

Prospective teachers often form images of science with sources of information other than firsthand experiences with science—including images in the popular media. Have your students go on a "scavenger hunt" to collect images of science and scientists. Have them share and critically evaluate the messages about science communicated in these images.

Science Autobiography

The Science Autobiography (Koch, 1990) is another early-semester assessment we have used to get a sense of our students' K-12 science experiences, as well as how they currently view science. In-class sharing sessions can serve as both pre-writing exercises and "get to know you" activities. For example, we begin by having each student think of a science memory, from elementary school if possible. Each student introduces him/herself and shares the memory. As instructors, we follow up with probing questions (What stands out to you most about that activity? How did that impact your feelings about science?), and survey the class to find out similar/different experiences (How many of you also...?). After each student has had a chance to share, we group students and give them the task of identifying common themes and patterns to their experiences. Each group generates a list on a whiteboard and then shares these with the group. This processing thus serves as a model for the kind of analysis and reflection emphasized in the Science Autobiography assignment.

Reading through students' science autobiographies, we develop a rich understanding of students' perspectives as science learners—the kinds of things they learned, the activities they were exposed to, and the kind of teaching they have had modeled for them. Specific episodes can be noted and referenced during future class sessions, to help methods students link course topics to their own experience. For example, in discussing assessment of science learning, it can be useful to draw attention to experiences students have had with traditional (i.e., paper-and-pencil test) and alternative assessment (such as projects, writing, and presentations). Additionally, we find that this early assessment gives us a good indication of students' writing ability (organization, expression, mechanics, etc.) and the level of reflection in which they engage. This helps us know where to provide additional support and modeling of our expectations for future assignments.

Lesson Planning Task

One of the simplest ways to find out methods students' ideas about science teaching is to ask them to write a lesson plan. We keep this task fairly open, allowing students to choose both the science topic and grade level. Note: we have found, despite explicit instructions not to do so, prospective teachers may rely on outside sources for ideas for planning their lesson. Therefore, we prefer to do this as an in-class writing assignment. Depending on the length of your class sessions, however, it may be more feasible to assign this as an out-of-class activity.

Students' initial lessons can provide you with an indication of the particular content/topics they think are appropriate for various age students, the level of understanding they have about a particular concept, their ideas about science pedagogy, as well as the kinds of information they feel should be included in a lesson plan. This can help you make adjustments to future assignments in which you ask students to plan instruction by being more specific about your expectations and providing additional supports such as model lessons. If you ask students to share how they chose their topic/lesson activities, you may find that they based their ideas on actual lessons they had experienced as learners. This can provide an opening for discussing the idea that "teachers tend to teach they way they were taught."

One advantage to using this pre-assessment is that students can compare their responses to the same task later in the semester and provide an indication to the instructor of the progress they have made in the methods course. Having students revisit their first lesson plan later in the semester can also make students' progress salient to them—often our students find much humor in their original ideas about science and science teaching, and are able to critique their previous lesson.

> ### It's Your Turn
>
> Try one of the above strategies, or one of your own, to find out your students' incoming ideas. *What do you see as potential points of resistance to learning to teach science among your students? How do you think you could use these as meaningful starting points for instruction?*

Conclusion

At the beginning of the chapter, we introduced you to a common point of resistance among prospective elementary teachers—that many expect that learning to teach science is simply a matter of learning a script to follow. When this belief is not challenged head-on, prospective teachers who are not provided with a script by their methods instructors may feel that they are not learning to teach. By getting to know your students' ideas about science and science teaching, you can more effectively plan instruction to meet their diverse needs, interests, and abilities.

> It was the third week of the semester, and I was joining my colleague Jan for lunch. "I decided to have my students write their science autobiographies," she began, "and I was really surprised to learn that my students had really positive feelings about hands-on activities. I

noticed, however, that they focused a lot on the 'fun' aspects of the activities they experienced in science classrooms, without really attending to what they learned."

"What did you change about your teaching, then?" I inquired.

"Well, I created an activity to do with them that compared what they learned from different kinds of activities—all of which were hands-on. This helped show them that not all hands-on activities are minds-on."

"And how did they respond?"

"They loved it—after we debriefed the experience as learners, focusing on what new understandings they developed, we debriefed what they learned from the activity as teachers. I can tell they gained a lot from it."

"It sounds like they could tell, too!" I applauded.

More to Explore

Below are several readings that provide examples of tools and tasks that you can use to assess your students' incoming ideas about science and science teaching. We encourage you to explore these as a supplement to the information provided in this chapter.

Cobern, W., & Loving, C. (1998). The card exchange: Introducing the philosophy of science. In W. McComas (Ed.), *The nature of science in science education: Rationales and strategies* (pp. 73–82). Dordrecht, the Netherlands: Kluwer.

Finson, K. (2002). Drawing a scientist: What we do and do not know after fifty years of drawings. *School Science and Mathematics, 102*, 335–345.

Friedrichsen, P., & Dana, T. (2003). Using a card sorting task to elicit and clarify science teaching orientations. *Journal of Science Teacher Education, 14*, 291–301.

Lederman, N. G., Abd-El-Khalick, F., Bell, R. L., & Schwartz, R. (2002). Views of nature of science questionnaire: Toward valid and meaningful assessment of learners' conceptions of nature of science. *Journal of Research in Science Teaching, 39*, 497–521.

References

Abell, S. K. (2007). Research on science teacher knowledge. In S. K. Abell & N. G. Lederman (Eds.), *Handbook of research on science education* (pp. 1105–1149). Mahwah, NJ: Lawrence Erlbaum.

Akerson, V. L., Flick, L. B., & Lederman, N. G. (2000). The influence of primary children's ideas in science on teaching practice. *Journal of Research in Science Teaching, 37*, 363–385.

Appleton, K. (1995). Student teachers' confidence to teach science: Is more science knowledge necessary to improve self-confidence? *International Journal of Science Education, 17*, 357–369.

Bryan, L. A., & Abell, S. K. (1999). The development of professional knowledge in learning to teach elementary science. *Journal of Research in Science Teaching, 36*, 121–139.

Chambers, D. W. (1983). Stereotypic images of the scientist: The Draw-A-Scientist Test. *Science Education, 67*, 255–265.

Gunstone, R. F., Slattery, M., Baird, J. R., & Northfield, J. R. (1993). A case study exploration of development in preservice science teachers. *Science Education, 77*, 47–73.

Hubbard, P., & Abell, S. (2005). Setting sail or missing the boat: Comparing the beliefs of preservice elementary teachers with and without an inquiry-based physics course. *Journal of Science Teacher Education, 16*, 5–25.

Jarrett, O. S. (1999). Science interest and confidence among preservice elementary teachers. *Journal of Elementary Science Education, 17*(1), 49–59.

Koch, J. (1990). The science autobiography. *Science and Children, 28*(3), 42–43.

Lederman, N. G. (2007). Nature of science: Past, present, and future. In S. K. Abell & N. G. Lederman (Eds.), *Handbook of research on science education* (pp. 831–879). Mahwah, NJ: Lawrence Erlbaum.

Magnusson, S., Krajcik, J., & Borko, H. (1999). Nature, sources, and development of pedagogical content knowledge for science teaching. In J. Gess-Newsome & N. G. Lederman (Eds.), *Examining pedagogical content knowledge: The construct and its implications for science education* (pp. 95–132). Boston: Kluwer.

Moore, J. J., & Watson, S. B. (1999). Contributors to the decision of elementary education majors to choose science as an academic concentration. *Journal of Elementary Science Education, 11*(1), 39–48.

Ramey-Gassert, L., Shroyer, M. G., & Staver, J. R. (1996). A qualitative study of factors influencing science teaching self-efficacy of elementary level teachers. *Science Education, 80*, 283–315.

Schoon, K. J., & Boone, W. J. (1998). Self-efficacy and alternative conceptions of science of preservice elementary teachers. *Science Education, 82*, 553–568.

Southerland, S. A., & Gess-Newsome, J. (1999). Preservice teachers' views of inclusive science teaching as shaped by images of teaching, learning, and knowledge. *Science Education, 83*, 131–150.

Tilgner, P. J. (1990). Avoiding science in the elementary school. *Science Education, 74*, 421–431.

Weiss, I. R., Banilower, E. R., McMahon, K. C., & Smith, P. S. (2001). *Report of the 2000 National Survey of Science and Mathematics Education*. Chapel Hill, NC: Horizon Research, Inc.

Young, B. J., & Kellogg, T. (1993). Science preparation and attitudes of pre-service elementary teachers. *Science Education, 77*, 279–291.

Chapter 5

Curriculum and Resources for Elementary Science Teacher Education

> "Noeline!" It was a phone call from a colleague, Mandy, from a nearby college. She sounded desperate. "The Dean has decided we will be doing elementary as well as secondary science methods. I've been assigned the elementary science methods, and only a month to pull it together."
>
> Noeline began to offer congratulations, but Mandy cut her off. "I haven't got a clue what to do, Noeline. Elementary majors are different from secondary. I suppose we will look at the curriculum, but we can't do that the whole time. We should do some science, but I know some elementary folks might have trouble with that … I was also told that a science lab in another building is open during the time I will teach the course. I'm not sure whether I should use the general purpose classroom I currently have scheduled, or switch so I'd have real lab tables, sinks, gas, a hood and more. Do elementary teachers even use that stuff?"
>
> "Why don't we get together and have a talk about it," Noeline suggested. "I can show you what I do, what resources I use, and help you develop your syllabus."
>
> "Thanks!" was the relieved response.

Every elementary science methods instructor faces the question that Mandy was wrestling with at some time or another. A course inherited from a previous instructor may make it a bit easier, but almost certainly there would be parts that need changing. And when a program is revised, a different schedule and placement of the science methods course within the program almost certainly would require some adjustment to what happens in the science methods course.

A considerable difficulty for elementary science methods instructors is the lack of information about what should be included in a methods course: there are no official guidelines, only the occasional publication that includes a report on another instructor's ideas, and the ubiquitous science methods textbooks. Too often it is these textbooks that define the science methods curriculum, yet one of the contributing areas of an instructor's PCK for teaching elementary science methods PCK is knowledge of curriculum. In this chapter, we consider some general principles about the curriculum for elementary science methods courses. Local conditions will result in greater or lesser emphasis of some aspects, about which you the reader will have to make judgments. There are also important contextual considerations, such as the emphasis on science as Inquiry.

The main components of curriculum that we will consider in this chapter are aims and goals, course content, the sequence in which to deal with topics within the course content, subject matter integration, and selecting resources. The chapter will conclude with a discussion of communicating expectations regarding the course through creation of a course syllabus.

Aims and Goals for the Science Methods Course

The main aim of a science methods course is to produce graduates who are ready to commence teaching science. That is, they have a "starter pack" of PCK for science teaching. Our science methods goals therefore focus on the development of this PCK starter pack. We suggest eight key goals. In a science methods course, prospective teachers:

- develop a working knowledge of the science curriculum;
- extend their understanding of selected science concepts;
- extend their understanding of the nature of science;
- develop a repertoire of strategies to investigate students' prior knowledge in science, and identify some commonalities in ideas;
- develop the knowledge and skills to design and implement sequences of science learning experiences that constitute inquiry investigations, are informed by students' prior knowledge, provide a scaffold to take students to the desired learnings, and cater for individual difference;
- plan for assessing student learning in science;
- collect and use multiple sources of evidence to make informed judgments about the progress of learners and the next steps in instruction;
- increase their self-confidence to teach science.

Other goals may be included depending on the context of the science methods course. For instance, if the methods course precedes the educational psychology course, then some goals related to constructivist views of learning would be necessary so the teachers could understand about the influence of prior knowledge on subsequent learning. Similarly, if the methods course precedes any in-school experience, the prospective teachers' views about schooling, their images of self as teachers, and their knowledge of general pedagogy would be very limited; in which case goals related to general pedagogy may need to be included. We therefore suggest that the science methods course should ideally not be scheduled in the first year of the elementary education program.

> **It's Your Turn**
>
> *How do these goals align with your goals for students? What additional goals do you have for them? How do these goals currently influence how you plan or design your methods course?*

The Methods Course Curriculum—What to Include?

Since the length/duration of science methods courses varies considerably between institutions, no definitive list of content to be addressed in a methods course[1] is possible. What is included (as well as what is excluded) from the curriculum of the methods course reflects a value judgment on the part of the methods instructor. How the content areas are dealt with depends on the instructor's orientation to teaching teachers (see Chapter 3). Though there is no universally agreed-upon set of ideas and skills one must address in the methods courses, the notion of PCK around which this book is organized provides us with a useful framework for considering the needs of our prospective teachers. We suggest a core set of content areas that would fully occupy most methods courses. These include:

Knowledge of Curriculum

The Official Elementary Science Curriculum

Prospective teachers need to be familiar with the official curriculum with which they are most likely to work when they graduate.

---

1. Listing content areas does not necessarily imply a "topics" orientation as outlined in Chapter 3. It is an essential component of course planning and ensuring alignment with other courses in the program.

Other National and State Policy Documents/Statements

Such policies and statements provide a context for the official curriculum, and shape its interpretation and implementation. Prospective teachers therefore need to be aware of these and how to operate within their requirements.

Knowledge of General Pedagogy

Teaching Science by Inquiry

Since this is a major issue in curriculum and policy documents, especially in the United States, it needs to be specifically addressed. Aspects to be considered include teaching models, teaching strategies and techniques, and routines.

Knowledge of Students

Students' Prior Knowledge

Prospective teachers need a firm grasp of the likely misconceptions held by elementary students about various science topics, and how their prior knowledge influences their learning.

Catering for Difference

A certainty for our prospective teachers is that they will encounter elementary students with diverse needs, interests, and abilities, so they need specific strategies for dealing with this in science. The minimum areas of difference that should be addressed are gender, social/ethnic, and physical.

Orientation to Teaching and Learning

Applying Theories of Learning to Science Teaching

We cannot assume that the prospective teachers commencing our elementary methods course have views about learning and teaching that are aligned with our expectations. We find that it is necessary to explicitly point out how aspects of, for example, constructivist views of learning are applied in devising scaffolded sequences of learning experiences (such as the teaching models outlined in Chapter 6).

Assessment

Planning and Implementing Assessment Strategies

Our prospective teachers need assessment strategies that will provide reliable assessment data that are derived from a variety of sources, especially in countries with a heavy emphasis on formal assessment and testing. They need to see that assessment can be built into the normal teaching regime rather than being an afterthought tacked onto the end of a unit of work. That is, it can be formative as well as summative (see Chapter 7).

Classroom Management

Managing Distribution and Collection of Equipment

Unless prospective teachers can manage this well, they have little hope of engaging elementary students in hands-on work.

Managing Small Group Work

This is common to other subjects, but needs to be highlighted in science as well.

Managing Field Trips

Field trips form an occasional component of learning experiences during an inquiry investigation. Prospective teachers need to be familiar with some key aspects of organizing field trips and managing behavior during these events.

Safety

Science activities sometimes involve potentially hazardous materials or tasks. Generally these should not be undertaken at elementary level, but authors of activity ideas do not necessarily consider this. Nonetheless, even fairly simple activities may require protective measures, such as use of goggles, to reduce the risk of injury. Prospective teachers need to evaluate each activity in terms of risk and safety, and be able to take appropriate precautions to protect students. Many states have legal requirements for risk assessments that need to be highlighted, as well as guidelines regarding the ethical and responsible use of animals in the science classroom. Prospective teachers should be familiar with such guidelines and where to find them.

Curriculum and Resources 85

Resources

Finding Equipment Resources

Prospective teachers often come into a methods course thinking that they will need to use complex laboratory equipment. They need help in recognizing that everyday equipment is sufficient in almost all elementary science experiences (exceptions are good quality magnets, microscopes, and some electrical circuit equipment). Some elementary schools invest in commercially available equipment packages, so prospective teachers need to be aware of those available.

Managing Equipment Storage

Having equipment is no good if it cannot be found or accessed easily when needed. An effective storage system is therefore necessary. This may depend on the school, but a new graduate's voice may influence school procedures.

Finding Curriculum Resources

Some state elementary science curricula include teacher and/or student resources, but many do not. Published and online resources are numerous. Teachers need to be able to find what they want quickly, and be able to evaluate what they have found.

Using Information Technology Resources

Nowadays, technology is ubiquitous and many classrooms (though sometimes lagging behind the latest advances) are equipped with a variety of technological tools to support instruction. These may include computers (whether in a lab or the classroom itself) equipped with the Internet, probeware, and other "hardware" as well as a variety of software packages or web-based applications. Prospective teachers need to be familiar with such tools, but also develop an understanding of how to best leverage such tools to support student learning.

Science Content Knowledge

Science Concepts Associated with Selected Science Topics

In most methods courses there is only sufficient time to deal with three to six science topics. It is preferable to deal with topics in depth, rather than cover many superficially. Topics common to the elementary curriculum should be chosen, but activities and concepts should be aimed at the capability of the prospective teachers. Physical science topics should be given preference, then earth/astronomical sciences, and finally biological sciences. This is a deliberate choice because prospective teachers' confidence ranks from biology (highest) to the physical sciences (lowest). Further, success in the area of science that they fear most provides confidence that they can learn other science topics successfully (Appleton, 1995).

Aspects of the Nature of Science

Most prospective teachers have misconceptions about science and its nature. Aspects of the nature of science therefore need to be explicitly addressed in the science methods course (Lederman, 2007). Prospective teachers need to develop an image of science consistent with the reforms and learn strategies for addressing elementary students' ideas about the nature of science as well.

Activities that Work

Planning and Implementing a Unit of Work in Science

We have found that prospective teachers need to design units of work in the supporting environment of the methods class with assistance from the instructor. Further, they need the opportunity to teach their planned unit.

> **It's Your Turn**
>
> This list of content areas is what we consider important for inclusion in an elementary science methods course.
>
> *Are there any of these content areas that you do not think important for your methods class? Why?*
> *Are there any other content areas that you think should be included? Why?*
>
> In the Activities that Work, we have selected particular science topics and concepts and several aspects of the nature of science. Compare our list of science topics and concepts, and aspects of the nature of science with those you think are important.

Planning the Curriculum

When you look over all the types of knowledge that prospective elementary teachers need, the first thought is a natural, "There's no way you can cover all that in a methods course!" That is, in fact, true if these are viewed as isolated bits of knowledge or topics to be covered in turn. Recall in Chapter 3, however, that there are multiple ways to organize a methods course, depending on your orientation toward teaching teachers. The theories of learning to which we subscribe, our own experience, and research show that it is preferable to teach in an integrated manner, versus organizing a course into a series of disconnected topics. And, in particular, it is important to teach science content from a student learning focus. So, what would such an integrated focus look like?

First, assuming that a typical class session is 3 hours, here are some essential components of each class session:

- time for methods students to ask questions;
- small group discussion of ideas from the text and other sources (like the online readings);
- reporting to the whole group;
- brief period of journal writing (i.e., the students keep a learning journal);
- spot "lectures" on key ideas; and
- participatory modeling of selected aspects of science teaching and learning, using selected science content. Showing a model unit plan and describing its construction has proved valuable for students.

In planning an entire course, we have found it helpful to use an organizing matrix that includes some of these major ideas (see "Tools" section). From organizing this semester-long calendar, the instructor moves to planning each class session. On the following pages, we include a representative schedule for 2 weeks of the methods course.

Sample Schedule, Week 1

1. Introductions.
2. Your own experiences of school science (pair; share with small group; share with whole group).
3. Course syllabus and textbooks.
4. Online resources.
5. Show the elementary science curriculum.
6. Give the moon assignment (see ATW 3).
7. Group roles using example activities:
    a. Distribute badges and assign roles (equipment manager, reporter, group director—see ATW 2).
    b. Provide a task using the orient–enhance–synthesize (OES) (ATW 5) and the Predict–Observe–Explain (POE) (see Chapter 6):
       Example OES–POE Task
       Orient. POE task done in small groups: fill a tumbler with water

till it is level with the brim. Predict how many washers can be added (carefully) to the tumbler before the water overflows. Observe—add washers one by one (fair test—same person in the same way). Explain your observations. Compare observations and explanations.

Enhance. POE task: fill a tumbler with water. Place a piece of card over the tumbler. Predict what will happen when the tumbler is turned upside down, holding the card in place, and then slowly releasing the card while upside down. Observe (do this over the sink or a basin). Repeat varying conditions (controlling variables), e.g., the quantity of water. Explain your observations. The explanation must be consistent for both activities.

Synthesize. POE task: place water in a pie pan to approximately half an inch deep. Predict what will happen when ground pepper is sprinkled onto the water. Observe. Explain (an explanation consistent with the previous activities is required). POE: predict what will happen when a drop of washing up detergent is placed in the center of the pan of water. Observe. Explain. A few students will understand about water tension, but many will not. Provide an explanation, and the effect of detergent. Use a drama model to help explain this[2] (see also modeling in ATW 5). Also point out that many texts will erroneously attribute the first activity to air pressure. Emphasize the need to treat sources cautiously.

8. Explain the respective components of the elementary science curriculum, pointing out how the activity conducted fits.
9. Revisit key points of the methods course.
10. Journal time.
11. Set reading tasks (on research into students' prior knowledge) and learning projects.

---

2. Selected students act as water molecules in one part of the room, with hands outstretched. They move constantly, briefly holding the hand of nearby molecules (simulating H–H bonding in water). Then have other students in an adjacent part of the room act as air molecules that are more spread out, move faster, and do not interact. At the air/water interface, water students can only hold hands with those beside and below them, simulating the surface forces of surface tension. This can then be illustrated on the board. Other students simulating detergent molecules can also be added. Water prefers to hold hands with one hand of the detergent molecule rather than other water molecules. The other hand of the detergent molecule prefers to touch air molecules, so there is less water–water hand holding at the air/water interface. That is, the surface tension forces are reduced. See ATW 5.

Sample Schedule, Week 2

1. Invite questions arising from last week. (The focus for this week is on the application of learning theories.)
2. Distribute assessment requirements.
3. Group discussion of readings from previous week.
4. Select four students to share their moon observations.
5. Workshop: all lessons and units of work have a start, development, and conclusion. What we do in each should be decided, in part, by ideas about learning.
6. Brainstorm ideas about how to start a unit of work, based on ideas from learning theory. List on board in two columns: strategy and reason/idea it is based on.
    a. Rule of thumb: never start with a video unless it is a vicarious experience in place of a real one, e.g., seeing a volcano.
    b. Second rule of thumb: choose something to "suck them in" as a motivator.
    c. Show examples for doing this such as Concept Cartoons CD (Keogh & Naylor, 1997), and discrepant events. Demonstrate a discrepant event: place a deflated balloon on the neck of a glass bottle. When the bottle is immersed in hot water the balloon slowly inflates until it is standing upright. Invite explanations, highlighting how this can provide insights into students' prior knowledge (relate to readings). Do not dwell on the explanations: most students will readily grasp the idea of air being heated and air pressure.
7. Brainstorm in a similar way strategies important in conducting a lesson, e.g., wait time, linking to what is known, scaffolding ideas, summarizing, modeling, asking questions, different types of questions. Write these on board with associated reasons/ideas.
8. Brainstorm ideas for the development of a unit: sequencing learning experiences and teaching strategies, e.g., being clear about what is to be taught and developing a sequence of logical steps in developing that idea; using evidence; building in the notion of challenge (Baird & Mitchell, 1987); challenging students' ideas (comparing different ideas, using confronting evidence, discrepant events, demanding consistent explanations across events).

    Example
    The idea that air exerts a pressure, introduced by the balloon on a bottle discrepant event (6c). Ask, "Where does air pressure come from?" (Someone will answer: the weight of the air.) Ask, "Does air really have weight? Is it a real substance?" State, "Real substances occupy space (volume) and have mass (weight)." Brainstorm for activities that might provide evidence that air occupies space, and has mass.

Select suitable ones and conduct them. For example, place a crumpled paper tissue in the bottom of a glass, and then place the inverted glass in a bowl of water. The air pocket should be visible, and the tissue should remain dry. The students will probably not think of an activity for mass, so this will need to be provided: construct an improvised beam balance by suspending a length of wooden dowel in the center by cotton thread (N.B. avoid breezes). If students do not understand how beam balances work, use masses to show how the beam remains balanced when equal weights are placed equidistant from the center. Attach two inflated balloons at either end of the dowel. Make minor adjustments with modeling clay to get the beam balanced. After obtaining student predictions, deflate one balloon by pricking near the neck (do not pop the balloon). The end of the beam with the inflated balloon will sink. That is, the air in the balloon has weight. Review the argument and the evidence for the conclusion that air exerts a pressure arising from the weight of the air.

9. For concluding a unit of work, suggest that it is best to use a new situation/problem that allows students to apply their knowledge. E.g., remove the lid from a quart plastic bottle, and immerse it in hot water. Leave for a few moments, screw the lid on tightly, remove it from the water, and place it on the table. After a few moments the bottle will collapse.
10. Review the specific application of learning theories used in the session. Also highlight other aspects such as the use of evidence (nature of science).
11. Journal time.
12. Set reading tasks (on the nature of science) and learning projects.

From these example sessions, it should be apparent that numerous aspects of pedagogy, concepts about the nature of science and science content, application of learning theories, student prior knowledge, and curriculum have been integrated. However, they are not implicit. Instead we make students aware of these aspects at appropriate times during the sessions. During the conduct of a simulated lesson, the prospective teachers are acting as students. We find it best to pause using some signal (e.g., make the comment, "An aside, from the teacher's viewpoint" accompanied by a sideways step, or ask the methods students to put on their "teacher-hat") to help the teachers make the mental transition from being student to viewing things from a teacher perspective. After the discussion and review of the teacher behavior, use a signal for the teachers to revert to student mode (e.g., a sideways step in the other direction with the words, "Back to the activity," or "Off with the teacher-hat and on with the student-hat").

We have found that an essential component of the methods class schedule is for the instructor to model, working cooperatively with the prospective teachers, the planning of a unit of work in science. If at all possible, we then assign the teachers, in pairs, to a small group of children (this may necessitate a series of visits to a school if a field experience is not concurrent with the methods course). The teachers, in their pairs, prepare a unit of work (this is best negotiated by the instructor with the children's teacher) spanning three or four lessons. The instructor confers with each pair about their plans, helping as necessary. The teachers then implement their plans with their small groups of children. Finally, they prepare a report on their teaching, including a comment on the children's learning that is evidence based. The unit plan and report can readily form part of the assessment requirements for the prospective teachers (see Chapter 7).

> **It's Your Turn**
>
> In this section we have emphasized the integration of different areas of knowledge.
>
> *To what extent do you integrate in your methods course?*
> *Consider how you might adjust your schedule and delivery to include greater integration.*
>
> We have also emphasized the need to make aspects of pedagogy explicit, and suggested that you need to interrupt the prospective teachers' "student mode" when doing this.
>
> *Are there ways in which you could make aspects of your pedagogy more explicit to your prospective teachers? How would you do this?*

Selecting Resources and Materials

The Methods Textbook

There are numerous resources available for teaching the elementary science methods course. The most obvious are textbooks. An early decision when designing a science methods course is whether to use a textbook, or whether to use a collection of resources. The former choice is convenient, but you are constrained by the author's choice of content and organization, which may or may not align well with your own preferences. The latter allows your course to be tailored very specifically to your own preferences, but takes some time to assemble. A third possibility is a compromise—use a textbook supplemented by other resources.

Most publishers will provide you, free of charge, with a review copy of a text for you to examine before deciding to use it in your course. This can allow you to determine which best meet your needs, as well as the needs of your students. We suggest you use several criteria when selecting a textbook:

- The extent to which the book aligns with the aims and goals of your course. This seems obvious enough, but different methods textbooks take markedly different approaches, based on the orientation of the author(s) toward teaching prospective teachers. Your textbook should align with your orientation, and should assist you in achieving the goals you've outlined for the course.
- The extent that it is used by others. If a book is popular, there is a good reason for it. However, a popular textbook is not necessarily the best textbook. Ask colleagues why they have chosen to use a particular book to better understand how the book aligns with your needs as an instructor.
- The author(s) experience and credibility. Authors with established reputations in elementary science education can usually be relied upon to have made wise choices about what to include in a book.
- The appeal to prospective teachers. This includes the layout and presentation, activity ideas, readability, relation to classroom practice, and cost. Students tend not to use a book that has little or no appeal to them.
- The utility of the book as a reference beyond the methods course. Students often sell textbooks back to a bookseller at the conclusion of the semester. We find value in utilizing textbooks that students will find helpful references during their teaching, and encourage students to retain them.

If you choose a textbook, ensure that your methods students have to use it to be successful in your course. Students tend to become annoyed if they purchase a book to which their instructor rarely refers.

Other Non-Textbook Materials and Resources

As indicated above, methods instructors may decide to forgo a textbook altogether or use one in conjunction with other non-textbook materials and resources. This can include both print and online materials. Some textbooks will include further references to both print and online material that you might want to investigate. Alternatively you can conduct your own search. We find it is wise to always check the resources personally before including them in a list of readings. This should be done at the start of each new semester: online resources in particular can be unreliable. Supplementary reading material has to be readily accessible to your prospective teachers. Most institutions use some form of online support or electronic reserve system that you can use, but the resources you choose must be in a suitable electronic format. Your methods students also need to be provided a list of the resources, how to access them, when to read them, and how they will be used in your pedagogy.

Just as there are advantages to using a textbook, there can be advantages to using non-text materials in your course. For example, one criticism we've heard raised by prospective teachers is that the textbook appears to provide only one author's perspective on teaching (whether or not this is actually the case). Thus, an advantage of using readings by various authors is that it can assist prospective teachers in considering science teaching and learning from multiple viewpoints. In particular, we have found that practitioner articles written by classroom teachers resonate well with our students—they appear to have "street credibility" in terms of illustrating how various pedagogies might play out in the actual classroom. Furthermore, we have found that many such articles (e.g., from the National Science Teachers Association elementary journal *Science and Children*) provide important background and connections to learning theory. That is, these can be an important resource for helping methods students develop their PCK for elementary science teaching by linking their knowledge of learning to their knowledge of pedagogy.

There are several questions you can use to guide your selection of non-textbook readings and resources:

- How will you include the reading in your schedule and pedagogy: is the reading a primary source for your teachers, or a supplement?
- How well does it fit with your list of content areas that are to be included in your course?
- Will there be some part of your schedule and pedagogy that monitors what the teachers have read, and provides opportunity for them to discuss and analyze it?
- Are you duplicating what has been said elsewhere in your course? This is appropriate if you are deliberately using the reading so your teachers can engage with the same information several times (Nuthall, 1999, 2001), but otherwise may be superfluous.
- How pertinent is the reading to the content or information that you want your teachers to engage with? If the reading deals with the topic obliquely it may not be worthwhile.
- Who is the audience the piece has been written for? Research reports written for other researchers are not always the best format for prospective teachers, though they may sometimes be useful.
- Is the piece of a suitable length for your teachers to be able to read during a hectic semester?
- How readable is the piece?

> **It's Your Turn**
>
> Review the print/online resources that you use in your course in the light of the discussion here. *Are there any changes you would want to make? Why?*

Elementary Curriculum Materials

As new teachers, your students will likely be given a science curriculum and be expected to implement it in their classroom. In order to help build elementary teachers' knowledge of curriculum, we find it helpful to utilize examples of elementary science curricula in our methods courses. This can acquaint prospective teachers with high-quality examples of reform-based curriculum materials, as well as help them think critically about how to best utilize curriculum materials to support students' learning.

Two kit-based curriculum materials for elementary science available in the United States and Canada include *Science & Technology for Children* (STC) and *Full Option Science System* (FOSS). Both of these are research-based versus being developed strictly for commercial purposes. *Primary Investigations* and *Primary Connections* are the two well-known sets of curricular materials available in Australia, but these are limited to teacher guides and pupil books—there are no equipment kits included (though some commercial enterprises have assembled equipment). Your college's education library may already have a collection of such curriculum materials, or you may be able to request their inclusion in the library's collection. If so, you and your students can readily access these materials for use both in and outside of class time. If you do not have access to these materials, you may be able to obtain examination copies through the curriculum publishers. Alternately, a local school could temporarily provide samples of curriculum materials utilized for teaching elementary science.

Being able to experience a model lesson taught by the methods instructor or work as a team to implement a lesson and receive feedback from others (microteaching) can be valuable experiences for your methods students in terms of developing knowledge of curriculum. We have also engaged students in examining curriculum materials to better understand the rationale behind the design of the curriculum, the resources included, and the organization of instruction. Methods students can use curriculum analysis tools developed by professional organizations (e.g., American Association for the Advancement of Science, no date; National Research Council, 1999) to analyze and critique the curriculum materials.

Facilities and Equipment

The Methods Classroom

As mentioned in Chapter 6, it is preferable for elementary science methods courses to be held in a general-purpose room rather than a science laboratory. One reason for this is to more closely model the environment in which prospective teachers will teach science. At the elementary level, science typically is taught in a multipurpose classroom, as opposed to a specialized science laboratory, as is usually the case at the middle and high school levels. Typical elementary school classes are self-contained, where a single teacher is responsible for teaching all or most of the academic subjects to a single group of students. At this level, flexibility of the classroom is key in terms of allowing enough space to accommodate a wide range of activities.

Having similar flexibility in the facilities used for your methods class is thus desirable, in terms of allowing you to engage your students in a wide variety of learning experiences, as well as helping create a comfortable learning environment. This comfort level is linked to student confidence and gender inclusive strategies. In terms of configuration of the room, we suggest avoiding the typical classroom arrangement of desks in a row, which sets up deliberate social barriers. It is easier to conduct small group discussions and workshops around a few desks put back to back, but larger group discussions are best done with seats in an approximate circle. If we plan to have a more formal presentation, then it is not hard to move the desks into a suitable configuration.

Equipment and Materials for Investigations

Just as with facilities, we also encourage you to utilize materials and equipment that are similar to those found in the elementary classroom. There are a number of reasons for this:

- Most prospective teachers are afraid of science, and the symbols that represent what they see as science include complex equipment. If they are faced with complex equipment, then there is potential that they may shut down emotionally.
- Our teachers' views of science often include misconceptions associated with science being done with complicated equipment. We need to specifically address this misconception.
- We want our prospective teachers to use equipment that they are likely to use in their own classrooms.
- Science laboratories where the equipment is stored do not suit the pedagogy that we want to use.

- Everyday equipment is inexpensive compared to science equipment, and elementary school budgets for science are usually quite limited. University budgets for science methods courses can also be limited, so this is an opportunity to model the use of inexpensive materials to our prospective teachers.

There are a few exceptions to this preference for using everyday materials. For instance, everyday equipment is sometimes not available for a particular purpose (e.g., an ammeter). However, you may find directions for constructing your own using inexpensive and readily available materials (e.g., paper cups, pins, and straws). Nevertheless, if the everyday equipment is of inferior quality (e.g., weak magnets, cheap microscopes with plastic lenses) it may not be worth using.

You should be able to find most needed equipment easily through a science equipment supply company that sells specifically to education buyers. Many science supply companies regularly mail catalogs to faculty, but you can also request free catalogs by visiting their websites. Comparison shopping is advised! Prices vary widely from supplier to supplier, and instructors should be wary of purchasing packages or sets of materials that appear to be cost-efficient, as it may actually be less expensive to purchase these items separately. Indeed, we've found that sometimes the same equipment can be more cheaply purchased from more general supply catalogs versus those marketed directly toward elementary schools.

In terms of obtaining everyday materials, most should be readily available in stores (and will be cheaper than if purchased through a science equipment supply company). A savvy instructor will also develop skills for seeing possibilities for the classroom during normal shopping trips. For example, strings of Christmas lights often go on clearance shortly after the season—these can be easily used for exploring electric circuits in the classroom, and purchasing them while they are deeply discounted can result in huge savings. Saving recycled materials (2-liter bottles, baby food jars with lids, etc.) is another strategy for stocking the science methods classroom (and the elementary classroom) cheaply. For example, the top of a 2-liter

> ### It's Your Turn
>
> Choose a science investigation or laboratory activity that you plan to do with your methods students. Make a list of all the equipment involved in the activity. *Which of these items would not be available in a typical elementary science classroom? What readily available substitutes could be used in their place? How would engaging your methods students in this same task assist them in envisioning certain activities as "doable" in an elementary classroom that has limited resources and/or funding for science materials?*

bottle can be cut off and used as a filter. The base of the bottle can be used as a waste station so that students can avoid having to travel back and forth to the sink and/or trash cans. Our point is to model best science teaching practices for prospective teachers, including finding appropriate equipment for science teaching.

Modeling Safety and Ethics

Though you may use everyday materials in a casual setting, it is important not to neglect the importance of safety and ethical considerations in conducting science investigations. For example, even in a fairly simple activity involving batteries, bulbs, and wires, a short circuit can cause wires to get extremely hot. Similarly, when exploring life cycles with mealworms, instructors should have a clear plan for what will happen to the beetles after the lesson or unit concludes.

Prospective teachers will be not only responsible for children's learning, but also for their safety in the science classroom. Helping methods students be aware of potential safety hazards and taking steps to prevent them is essential. They may view this as part of classroom management—such as establishing rules for how to investigate safely. However, we emphasize the importance of helping methods students understand why particular safety precautions are necessary, which can also be linked to understanding the content. For example, in the case of the short circuit, both methods and elementary students should understand that electrical current can produce both light (when a bulb is included in the circuit) and heat. (This is the reason for the plastic insulation on the wires.) When current travels from the battery, through the wire, and directly back to the battery without passing through another resistor, it can get very hot. Thus, understanding the safety in working with electrical circuits also has implications for safety in the home. Within this lesson, safety can also provide a meaningful real-world connection for students. Similarly, you may be working with live animals, such as mealworms (darkling beetles) in your methods course (see ATW 4). In this case, understanding the ethical responsibilities for working with live animals provides a connection to explicitly discussing the nature of science, and the ethical principles that guide scientific work.

It's Your Turn

Think about a science investigation you plan to engage your students in conducting. *What important safety and ethical considerations are there for doing the activity with young learners? How can you make these considerations an explicit part of the discussion and debriefing of this experience?*

Communicating Your Expectations to Students: The Course Syllabus

The syllabus (variously called a course profile, course outline, unit outline, or unit profile in some countries) essentially forms a contract between the instructor and the methods students as to what will take place during the course, and outlines both student/faculty roles and responsibilities. Your university may have a template or list of guidelines for creating syllabi, or you may be left to your own devices in designing your course syllabus. In general, we recommend the following components for your syllabus, whether it is online or on paper:

- information about the course (such as code/number, number of credit hours, prerequisites) and where it fits into the overall program of study;
- meeting times and locations;
- instructor contact information and office hours;
- summary/statement of the course purpose, aims, and expected outcomes;
- description of course content/topics to be addressed;
- required materials (textbooks, access to online resources, etc.);
- major assignments and their contribution toward final course grade;
- grading scales and policies regarding late work, etc.;
- attendance policy;
- other academic policies of your university (plagiarism, etc.) and expectations regarding work outside of class and study suggestions.

We suggest including expectations regarding attendance, studying, etc. in advance versus having to figure out how to address problems as they occur—be prepared to stand by any policy or expectation you include in your syllabus! We also suggest that, as part of your syllabus materials, you provide students with a description of each assignment and the evaluation criteria in advance, as well as an outline or schedule of course activities. Of course, it stands to reason that if you are flexible and responsive to students' needs, interests, and abilities, this schedule may be adjusted as you see fit—such a caveat should be included along with the schedule.

Conclusion

At the beginning of the chapter, Mandy was struggling with being new to teaching an elementary methods course. For instructors who do not have a background in elementary science teaching, this can be an especially daunting task. In this chapter, we considered the aims and goals of the elementary methods course, as well as considerations for the scope and content of the course. Our account of science methods curriculum and resources outlined in this chapter obviously reflects our own orientations to teaching teachers derived from many years of reading, research, and practice. We think it is significant that each of us holds a similar orientation given our diverse backgrounds. While we do not say that our orientation is "correct" and others incorrect, we suggest that you may find it helpful to honestly reevaluate your own views in the light of our suggestions.

A naïve perspective on curriculum and resources for science methods courses would be to dismiss these as unproblematic and straightforward—seeing the course merely as a syllabus, schedule, and resources. However, we suggest that these are just as important of a design feature as the pedagogy employed in your science methods classes. Much of the methods instructor's own views of teaching, learning, and science are revealed through the hidden curriculum using the windows of syllabus, schedule, and resources (as well as pedagogy). Just as we might overtly model aspects of what we want our prospective teachers to learn, we need to be conscious that we implicitly model other aspects through our orientation to teaching teachers and choice of curriculum and resources.

> It was the week before the start of the semester, and Mandy and Noeline were meeting to go over the syllabus she had been working on developing. "Once I had a clear idea of my goals," Mandy began, "everything else seemed to fall into place! I found more resources to use than I expected, but I realized that even though there are a lot of things that can be put into a course, it was best to stick to what would help students meet those goals."
>
> "So you have everything planned out, then?" Noeline replied.
>
> "Well—, not quite! I am sure that there will be things I need to tweak throughout the semester, and changes I might make to the schedule, *but*—, I have a clear vision now of the course and what I want students to know after they leave the course."
>
> Noeline applauded, "It sounds like you're in good shape for the start of the semester!"

Tools for Teaching the Elementary Science Methods Course

A Course Planning Matrix

Elementary science methods schedule

Week	Pedagogy content	Science content	NOS content	Reading	Journal writing	Equipment/ materials	Major assignments
1							
2							
3							
4							
5							
6							

More to Explore

Abell, S. K., & Bryan, L. A. (1997). Reconceptualizing the elementary science methods course using a reflection orientation. *Journal of Science Teacher Education, 8*, 153–166.

Kwan, T., & Texley, J. (2002). *Exploring safely: A guide for elementary teachers.* Arlington, VA: NSTA Press.

National Science Resources Center. (1996). *Resources for teaching elementary school science.* Washington, DC: National Academy Press.

The Association for Science Teacher Education (U.S.) recently undertook a "Syllabus Sharing Project." Examples of course syllabi for a variety of science education courses, including elementary science methods, can be found online: retrieved September 21, 2009, from http://aste.chem.pitt.edu/cgi-bin/syllabussharingviewing.pl. In addition, a quick Google™ search may yield other examples of syllabi.

References

American Association for the Advancement of Science (no date). *AAAS Project 2061 middle grades science textbooks evaluation: Criteria for evaluating the quality of instructional support.* Retrieved July 1, 2009, from http://www.project2061.org/publications/textbook/mgsci/summary/criteria.htm.

Appleton, K. (1995). Student teachers' confidence to teach science: Is more science knowledge necessary to improve self-confidence? *International Journal of Science Education, 19*, 357–369.

Baird, J. R., & Mitchell, I. J. (1987). *Improving the quality of teaching and learning: An Australian case study—the PEEL Project.* Melbourne, Australia: PEEL Group.

Keogh, B., & Naylor, S. (1997). *Starting points for science investigations.* Manchester, United Kingdom: Millfield Press.

Lederman, N. G. (2007). Nature of science: Past, present, and future. In S. K. Abell & N. G. Lederman (Eds.), *Handbook of research on science education* (pp. 831–879). Mahwah, NJ: Lawrence Erlbaum.

National Research Council. (1999). *Selecting instructional materials.* Washington, DC: National Academy Press.

Nuthall, G. (1999). The way students learn: Acquiring knowledge from an integrated science and social studies unit. *Elementary School Journal, 99*, 303–341.

Nuthall, G. (2001). Understanding how classroom experience shapes students' minds. *Unterrichts Wissenschaft, 29*, 224–267.

Chapter 6

Instructional Strategies for the Elementary Science Methods Course

> Noeline groaned as she glanced over the memo notifying her of her elementary science methods class schedule. Each class was listed as a 1-hour lecture and a 2-hour lab in the chemistry laboratory. She immediately visited Elias, the new administrative assistant responsible for the schedule. After initial pleasantries, she said, "This schedule won't work, Elias. It will turn the students off." "Why what's wrong with it?" he queried. "I thought you would want a lab and it is way under-used." "The chemistry lab will freak them out," she replied, "and I don't do lectures. Can I just have a 3-hour block in J47? It is big enough, has moveable desks, a few sinks, and some storage areas." Elias countered, "That'll be more work for you, with two classes: 6 hours instead of 5." Noticing the look on Noeline's face, he quickly followed with, "I'll see what I can do."
>
> Later, Noeline met with her colleagues, Sam and Joan, to review the science methods course material for the forthcoming semester. After going through the overview of the content for each session, Sam and Joan began to assemble their papers, thinking that they were done. But Noeline went on, "It is important to include a focus on some aspect of pedagogy in each session. That means both explaining it, and modeling it. The students like my way of categorizing pedagogy into routines, techniques and strategies, and teaching models. Are you familiar with those?" Sam and Joan looked at each other blankly.

The traditional way of teaching at universities and colleges is delivering a lot of information to large groups of students via lectures. These are followed up with small group tutorials, and in the sciences, laboratory work, where the lecture material is reviewed, practiced, and confirmed through experiments. This mode of teaching began in medieval days when information was mostly in the head of the Professor, who wrote the lecture notes that were then read to students by a Reader. Tutors in turn ran the tutorial groups where the information was revisited, and Demonstrators demonstrated to students the laboratory experiments referred to in the lectures. Many universities still organize classes in similar ways. University administrators all too often assume that a science methods class will operate the same way as the sciences. However, as Noeline mentioned, such an arrangement can be counter-productive, and is not a model conducive for effective student learning in the methods course setting. In particular, it draws on a form of pedagogy that is no longer considered appropriate for teaching prospective elementary science teachers. In this chapter, we explore alternative instructional strategies that can be used in elementary science methods courses, which are based on an instructor's pedagogical content knowledge for teaching science teachers.

Aspects of PCK Pertinent to this Chapter

A number of aspects of PCK for teaching science teachers relate to instructional strategies. These are highlighted below:

Orientation to Teaching and Learning

Elias's view of student learning was that students receive the information passed on in lectures, are shown supporting experimental evidence in laboratory classes, and learn the content at home. This view shaped his perception that the main instructional strategies should be lectures and laboratory experiments. This was clearly at odds with Noeline's views. Therefore, a critical underlying contributor to your PCK for teaching science teachers is your view of learning and teaching at tertiary level (see also Chapter 3).

Views about Assessment

Implicit in Elias's ideas about learning and teaching is that assessment should be via examinations. In Chapter 7 we explore a variety of alternatives.

Contextual Factors—Environment

Noeline rejected the idea of holding classes in a chemistry laboratory because she did not think this provided a suitable learning environment for prospective elementary science teachers. What constitutes a suitable environment is therefore a matter we consider in this chapter.

Knowledge of Students

The prospective teachers' own views about learning and teaching are important considerations in developing instructional strategies for science methods classes. If this is not taken into account, methods students may feel that they are not being taught anything and develop a dislike of the course and the instructor. Review Chapter 4 for perspectives on this.

Knowledge of General Pedagogy and Science Instructional Strategies

This has a dual focus, in that elementary methods instructors use their knowledge of pedagogy to construct PCK for teaching science teachers, but also help prospective teachers develop their own knowledge of general pedagogy and their PCK for teaching science. That is, instructional strategies have their origins in a teacher's knowledge of general pedagogy. When applied to the teaching of a specific science or science methods topic, they become incorporated into the PCK associated with that topic. Science methods instructors can readily draw upon their existing pedagogical knowledge, but the prospective teachers in their methods classes have little or no knowledge of general pedagogy to draw upon. Therefore, some aspects of general pedagogy necessarily become components of both the content and instructional strategies used in methods courses. For instance, during instruction about questioning, prospective teachers may become aware that open-ended questions should be used to facilitate inquiry (content: open-ended questions; instructional strategy: modeling how to use open-ended questions during an inquiry session on a specific science topic). However, it is when being involved with asking questions of school students during a science lesson on, say, igneous rocks that prospective teachers build into their practice the appropriate use of open-ended questions for science teaching. It is therefore preferable that in a methods course these instructional strategies be taught both generally and specifically through selected science topics: methods students can develop general pedagogical knowledge and PCK for teaching science conjointly and concurrently.

In the next section, we look more closely at those aspects of PCK for teaching science teachers not dealt with in other chapters.

Your Orientation to Learning and Teaching

Learning

The prospective teachers who are in our science methods classes are adults, and use adult ways of learning. Of course it is possible that those who have come to us straight from school may bring with them some previously acquired non-adult learning behaviors that may have been successful in school settings. Ways of learning, including those of adults, are explored in Chapter 1. If you read some books on adult learning, you will find very similar ideas, though the terminology used is sometimes different.

> ### It's Your Turn
>
> It is critical that you ensure that your views about how your prospective teachers learn are based on contemporary views rather than traditional views of university learning. Take a few moments now to reflect on this.
>
> *When you did your university degree(s), how did you learn?*
>
> Using the learning theories in Chapter 1 and the ideas about assessment in Chapter 7, develop some summary statements about adult learning. Compare these to how you learned at university yourself.

Our Answer

In our undergraduate courses we were provided with large amounts of information in lectures and texts. Our learning strategy was to work through this during classes and afterwards to try to understand the ideas and how they fit with what we already knew. In preparation for examinations we reviewed the ideas and spent considerable time writing and verbally reviewing both facts and ideas. We would also discuss the ideas with fellow students. At post-graduate level, this changed to a form of inquiry where we worked with an issue or problem, formulating ways of obtaining information about it, and evaluating the information. These we would discuss with our supervisor, all the time recording our ideas and investigations.

Adult learning is best centered on a challenge or problem. Students devise ways of accessing relevant information, evaluate it, and ensure that they understand the embedded ideas. This may entail individual work, but often includes discussion and observing others engaging in activities related to the issue or problem. In short, adults learn by engaging in inquiry. When something has to be learned by rote, they first develop an understanding of what they are learning before using repetition and multiple-sensory strategies to over-learn the material. Rote learning should play a minor role in science methods courses.

However, some young adults have the mistaken idea that learning is mainly about repeating facts many times until it "sticks." A consequence is that they avoid learning anything complex, expect to be told the facts to learn so they can be recalled in a test, and do not associate thinking or discussion with learning. In your methods classes it is likely that you will have some prospective teachers who hold this view. You may need to help them be metacognitive about what they are learning as the course progresses.

Teaching

Your orientation to teaching elementary science teachers (see Chapter 3) is related to your views about teaching as well as learning. Your views about adult learning provide pointers to how you go about teaching your methods classes and the instructional strategies you use. For instance, based on our views and our preference for assignment and project-based assessments, general instructional strategies that we use in our methods classes include creating a "need to know" by providing a challenge; providing a variety of information sources (including some "mini-lectures"); helping prospective teachers reflect on and evaluate the information and their learning; modeling aspects of science teaching, including instructional strategies; using small group activities, and implementing focused discussion in both small and larger groups.

> **It's Your Turn**
>
> Use your ideas about learning that you earlier wrote in your summary statements to make a list of general instructional strategies pertinent to your science methods classes.
>
> *Compare your list of instructional strategies to ours. To what extent are they similar or different?*

Environment

In our vignette at the beginning of the chapter, Noeline did not want to give lectures or hold her classes in the chemistry laboratory, as she did not consider this to be an environment conducive to learning for her students. Noeline's reluctance to conduct lectures should be evident from your review of learning theories, but her dislike of science laboratories may be harder to understand. This comes from her knowledge of her students. While some elementary science methods instructors may be fortunate to have a dedicated room in which to hold classes, in many institutions rooms are allocated to classes by demand and availability. Just like Elias, administrators may think that they are doing you, the methods instructor, a favor by putting your class in a science laboratory that is under-used. Unfortunately, as Noeline knew, many of her prospective teachers have developed a fear of science during their schooling (see Chapter 4). To come into a science laboratory would freak them out, and they would see themselves as failures in science methods before they even started. So, a more neutral room is preferable, as long as it has sufficient space and seating flexibility to enable both small group work and whole class seating. Some storage space in the room makes it easier for the instructor to have materials handy, and a faucet and sink can be very helpful.

General Pedagogy

General pedagogy is the range of instructional strategies that can be used to help students learn. When used in a particular context for a specific topic, the general pedagogy becomes transformed and embedded into the PCK for teaching that topic. As mentioned earlier, in a science methods course, the focus on general pedagogy is therefore twofold: (1) selected instructional strategies are part of the content to be delivered (and preferably modeled) in the course, and (2) many of these are also used during class in the delivery of the course. This happens especially when a selected strategy is taught by both information delivery and by modeling, as in an inquiry investigation (see Activities that Work section of this book).

Inquiry

Any discussion of pedagogy in a science education context needs to take into consideration inquiry, particularly within the United States. The decades old emphasis on inquiry in the U.S. was renewed in the Benchmarks for Science Literacy (American Association for the Advancement of Science, 1993) and the National Research Council's (National Research Council (NRC), 1996, 2000) standards. These documents aimed to engage school students with science in a similar way to that which scientists do, but unfortunately the message about inquiry has become confused and distorted. This is despite a fairly clear definition by the NRC (2000), to the effect that inquiry consists of the activities and thinking processes of scientists and knowledge of what scientists do, and what science is—all with a deep understanding of science concepts. The confusion and distortion has largely come from a few decades earlier when "discovery" was promoted as a pedagogical approach in science. Discovery was, and still is, often equated with inquiry. Associated with this view of discovery learning is the idea that the teacher should be a facilitator of learning, rather than a deliverer of information. From this perspective, the teacher is seen as playing a minimal role so the students can "discover" the science concepts—in particular, that the teacher should not tell students the answer. This view is encouraged by the use of the term "facilitator" in *Inquiry and the National Science Education Standards* (NRC, 2000).

Your colleagues may be some of those people who hold this view of inquiry, and many of your prospective teachers will certainly hold it. It may even be a perception you have yourself. We wish to state upfront that we believe this view is erroneous and unhelpful; that is, it is a misconception about science pedagogy that you will need to specifically deal with in your methods classes. Conversely, we emphasize that the nature of science is an important aspect of science learning for both school students and methods students, but that it needs to be taught explicitly and experientially. So from our perspective, inquiry needs to take a prominent role in pedagogy modeled for our prospective students, so they in turn can learn how to engage their elementary science students in inquiry and help them understand the nature of science.

This leads us to the issue of what inquiry in a methods class might look like. There are two considerations: inquiry as an orientation to teaching prospective science teachers (see Chapter 3), and inquiry as a pedagogical approach in science for both modeled science lessons and elementary science methods sessions. In this chapter, we focus on the latter. We have avoided calling inquiry an instructional strategy—it is more like a general pedagogical approach within which a number of instructional strategies may be nested. We will examine some of these instructional strategies later.

Regardless of the instructional strategy, it is important to recognize that it is essential that the teacher identifies what the students are expected to learn (goals), what they know already, and devise an instructional (learning) scaffold that will help them progress from where they are to the desired goals (with appropriate formative assessment strategies to monitor progress). Such an instructional scaffold consists of a series of different, carefully selected instructional strategies. This applies to both elementary science teaching and teaching prospective science teachers.

The notion of instructional strategies encompasses three types of strategies[1] that range in complexity: routines, techniques and strategies, and teaching models (which come closest to the notion of inquiry). However,

---

It's Your Turn

Jot down your own definitions of routines, techniques, strategies, and teaching models, with a few examples of each.
Keep these for reference as you read the rest of the chapter.

---

1. You may consider that our use of these terms is idiosyncratic, but we have discovered that prospective teachers find this categorization very helpful.

inquiry teaching models are dependent upon, and built from, routines, techniques, and strategies; so they all need to be considered. In the next section we explore several examples for each of these, as there are particular ones that need to be both taught and modeled in science methods courses, as well as used as instructional strategies in science methods classes.

Routines

Anybody who has been a part of formal schooling has learned hundreds of routines during that time. Thus is it likely that prospective teachers will enter your methods class aware of a large number of these routines. In fact, the routines have probably become so embedded into their perceptions of the way classrooms work, that they will not even be aware that they exist—they are just part of the classroom culture and are effectively invisible. On the other hand, their awareness of routines will be from a student perspective: they are used to the teacher initiating the routines and are not aware of the subsequent teacher–student negotiation about how the routines would work in a particular teacher's class. Consider the routine of raising hands in response to a teacher question. Students learn this within the first few months of schooling, but with each new teacher they negotiate the extent that the routine is flexible by engaging in other behavior such as calling out. After a few days in a given classroom, they know the different contexts when they are best to raise their hands and when they can get away with calling out answers.

In your first session with your methods class, if you demanded that they raise hands when they know the answer to your questions, then most will probably do so with the accompaniment of vocal and body language protestations (we would only suggest you do this as a pedagogical strategy to illustrate the nature of routines). In a recognizable classroom context, even in a college or university setting, they respond as students by following the routine, but using a more adult way of negotiating how the routine is used. Part of the challenge for you as instructor is to help them shift their thinking from student to teacher; from recipient of routine rules to the initiator of them. This means that you should draw their attention to the routine, and discuss with them its rules, applicable contexts, possible variations, and reasons for using the routine.

A few routines potentially applicable to your methods class follow. Your students could also adapt the routines for their own use in elementary science classrooms.

Classroom Seating

This may seem trivial, but it is important. Typical College and University room organizations are for desks to be in neat rows facing the whiteboard and screen. This is fine for lecturing, but you do not want to be doing a lot of lecturing; you want discussion and activity interspersed with the occasional mini-lecture. So, a routine to consider is to have the students move the seating into your preferred configuration on entry. Depending on the group size, the best arrangement for large group discussion is an open circle. For activity, desks should be clumped to form groups of about four students. Part of the routine may have to include replacing the furniture into its original configuration at the end of the class, if the space is shared with other instructors.

Why have a classroom seating arrangement to facilitate discussion and activity? This has at its heart our views of learning and teaching. We believe that both discussion and activity are important instructional strategies that enhance learning, aid reflection, and that also contribute to the development of appropriate views of the nature of science.

Classroom Transactions

These are the people-to-people interactions during your class sessions. The hand raising mentioned earlier is obviously not appropriate for adults, so what routine do you want for your methods class that will facilitate discussion? Consider the following:

When at the front of the class, replace the usual Teacher question—Student response—Teacher affirmation/rebuttal discourse pattern with Teacher question—Student response—Teacher wait (perhaps with body language to invite comment)—Student response—Student response. Waiting for maybe 30 seconds to a minute after a student response (sometimes called wait time 2—see the asking questions technique, below) can encourage discussion.

When using the open-circle seating pattern for discussion, start off the discussion, and then sit with the students in the circle in a non-prominent position. Follow the discussion with interest, but do not comment, even though they will start looking to you to affirm what is being said. Only say something when they have talked it through, preferably to ask for further elaboration, or for a thought to be followed up. At the end of the discussion, summarize the main ideas to provide closure. By sitting with them in a non-prominent position and not commenting, you are effectively abdicating the normal classroom teacher role. They will find this unsettling at first and try to make you assume a more prominent role. Some may even accuse you of not teaching anything!

Equipment Distribution

When the room is organized for activity, there will be clumps of desks with small groups seated at them spread throughout the room. To distribute materials to the groups, have each group appoint a materials manager. Place the materials at the front or side of the room, with items well spaced. The materials managers then come and collect the items. You may even wish to define a direction of movement. This avoids congestion at the collection point and the inevitable problem of two people from the same group collecting the materials, leaving one group without anything. It also models good practice for teaching elementary students.

Techniques and Strategies

In education, terms like techniques and strategies are ill defined and often used interchangeably, so we first outline what we mean when we use these terms. We see techniques and strategies as being slightly different, but related; both used during a lesson. Techniques tend to be single teaching actions that are common to most subjects, such as questioning. Although a single teaching action, a technique can be quite complex; for instance, questioning includes many facets, such as use of open and closed questions, distribution of questions in the class, and wait time. A technique forms part of an overall lesson scaffold for the students' learning, and serves a particular purpose in guiding learning. It is usually used in association with a routine or another technique. For example, an elementary teacher will often use the technique of asking questions of a class using a "whole class discussion" routine (where the students are seated in rows or on the carpet facing the teacher, bidding for a turn to answer questions by raising hands).

Strategies on the other hand are multiple teaching actions, usually consisting of a mix of routines and techniques. Some strategies can be used in different subjects, but often they fit best within a particular subject, so we will limit the examples discussed here to those suitable for science and therefore science methods. A lesson may include just one teaching strategy, sometimes two, or even three for a longer lesson. Occasionally a teaching strategy may spread over two lessons. An example of a well-known science strategy is Predict–Observe–Explain (POE). The purpose of a teaching strategy is to provide a scaffolding structure for the lesson. For example, when using POE, a teacher would usually use another strategy to provide an introductory focus prior to the POE sequence, such as a demonstration or a hands-on activity in conjunction with (say) a small group activity routine. After using POE, the lesson would be concluded with another strategy, such as writing a class summary on the board. Together, these strategies form a learning scaffold with an attention-getting and motivating introduction, a scaffold for encouraging students to arrive at their own explanations for an event, and a conclusion that summarizes the main ideas to be learned.

There are quite a few techniques and strategies that are important in science teaching, which should be included in the science methods curriculum. A useful instructional strategy for teaching these in your methods class is to explain the technique/strategy, model it using a science activity, and then deconstruct it. This way your students are introduced to it from a teacher's perspective, then experience it as a learner, and finally analyze and reflect on it as a teacher. Many of the techniques and strategies are also useful pedagogical tools for scaffolding science methods sessions. For instance, questioning and explaining an idea are techniques common to both science and science methods. A selection of techniques and strategies are briefly highlighted below, some of which feature in the Activities that Work section.

Explaining an Idea Technique

There is no one "right" way to explain an idea, but characteristics of a good explanation are that it is clear, focused on the topic, and brief (Smith & Land, 1981). Examples and analogies are helpful but not essential components. Similarly, visual representations can be helpful—provided that the representations that model the idea have links to the idea pointed out, and limitations of the model are highlighted.

Other principles are that the complexity and conceptual demand of the explanation should be tailored to the level of the audience, and the starting point for the explanation should be based on what the students already know about the idea being explained. That is, how to make the idea understandable to the students needs to be considered (an important aspect of PCK). It is also critical to time the explanation appropriately in the pedagogical sequence. Sometimes an explanation is best placed near the beginning of a session, but most often it is better placed after the students have been working with the idea themselves and need some clarification and closure. That is, a "need to know" has been created. Engaging in lengthy explanations without creating a need to know can result in, at least, inattentiveness; and at worst, behavioral problems—even in a methods class. Adults and even some children are quite adept at feigning attention while being mentally elsewhere.

Some people have the mistaken belief that if science instruction is to be inquiry based (NRC, 1996), then there should be no teacher explanations—a hang-over from the days when there was a strong belief in "discovery learning." We wish to make it quite clear that explanations form a critical component of inquiry learning, provided they are done within a carefully thought out scaffold for learning, are not done too often or are too lengthy, and are preferably introduced after a "need to know" has been established. Within a methods course, explanations should be included as part of a modeled inquiry session, as well as when (for example) explaining aspects of pedagogy and the bases for making pedagogical choices.

It is also salient to consider Nuthall's findings (1999, 2001) that grade 6 students need to encounter the same information at least three times in order to recall it subsequently. It is likely that the same could apply to adults. That is, just because an idea has been explained does not mean that the students have understood it or will remember it. Pedagogically, the instructor should devise three different ways for the students to engage with the idea, each occasion being within two days of the previous experience. This can be challenging in University and College contexts that are governed by rigid timetables, but does provide opportunities for including other types of learning such as solo or group project work, library research, school-based (field work) assignments, or online tasks.

Concept Mapping Technique

Concept mapping is a useful technique to investigate and/or clarify a student's understanding of a concept or idea (see also Chapter 7). Although concept maps can be used for summative assessment (Bell, 2007), they are best used diagnostically early in, or during, an instructional sequence. The use of concept maps in science has been well researched—see, for example, the special issue on concept mapping in the *Journal of Research in Science Teaching* (volume 27, 1990).

Concept maps can be used in methods courses to explore prospective teachers' ideas of science concepts, as well as educational concepts such as learning theories and their classroom application. We have also used them to outline a particular elementary science pedagogical sequence,[2] and then explored the reasons why that sequence provides a learning scaffold for elementary students.

The following steps provide a guide to constructing a concept map:

1. Identify the concept or idea for which you will construct a map.
2. Jot down a list of words (nouns) that you associate with the concept/idea. It is best to brainstorm these quickly. Using sticky notes (one word per note) or appropriate computer software (such as *Inspiration*, or for elementary students, *Kidspiration*)[3] makes it easy, but a pencil and eraser work fine.
3. Arrange the words in a logical order that makes sense to you. You could make a linear sequence, several sequences, or a web. Swap the words around until you are satisfied with the arrangement (sticky notes or software makes this easy).
4. Explain how each word is related to the one next to it by drawing a relationship arrow and putting on the arrow a word (usually a verb)

---

2. For instance, we first brainstorm activity ideas for teaching a particular science topic, then use concept mapping to organize the activities into a logical sequence that forms a learning scaffold. Links between the activities consist of instructional strategies.
3. *Inspiration* and *Kidspiration* are produced by Inspiration Software Inc.

or phrase to summarize the relationship. If you find a word out of place, move it. You may even decide that you need to change the whole organization. New words may occur to you as you do this, so add them into your concept map.
5. You will rarely be completely satisfied with your concept map, so stop when you think it is reasonable. As your ideas change over time, the concept maps that you construct for the same concept/idea may well change.

Once the technique of constructing a concept map is familiar, it can be readily embedded into a teaching strategy. For instance, it is often useful to have students share their concept maps with others, perhaps using a think-pair-share routine.[4] Explaining their maps to others forces them to rethink what they have done and identify any inconsistencies. Others' questions may also help them clarify aspects of their maps. For example, a teaching strategy, perhaps called "group concept map," could begin with the technique of students constructing their own personal concept maps, sharing them with a partner and creating a joint map, which is then shared in a bigger group (such as two pairs) that then creates one concept map for the whole group.

Asking Questions Technique

A lot has been discovered about asking questions that enhance learning, so just a brief outline of the important components is outlined here. First, questions can be used for different purposes: as a management strategy, such as keeping students on task; as a way to rapidly test students' recall of information; or as a means of encouraging students to think about the available information. The last of these is the most useful for enhancing learning, but is also not used as frequently as the others. Three issues for enhancing learning to consider are:

The Type of Question. Generally, questions can be categorized as closed (convergent) or open (divergent). Closed questions usually require a brief answer, and are often used in a rapid sequence; for instance, "What is the boiling point of water? Its freezing point?" These are best for recall of information and there is normally one correct answer. They should be used sparingly if the purpose is to enhance learning. Open questions have several possible answers, and encourage a greater depth of thinking; for example, "What do you think we could do to change the boiling point of water?" or "What teaching approach do you think would be effective to

---

4. In the think-pair-share routine, students each *think* of their own answer, then *pair* with another to share. The best idea is selected, and *shared* in the small group or whole class.

teach 'Floating and Sinking?' " This provides the students with the opportunity to raise several possible answers that could, if desired, then be tested (e.g., "What would happen if...?" teaching strategy for a science topic). Notice that the questions could have been phrased "How could we change the boiling point of water?" or "What is the best way to teach 'Floating and Sinking?' "; but we have found that the inclusion of the "you think" phrase further reduces the feelings of many students and prospective teachers that the instructor is looking for a specific correct answer. As elaborated in Chapter 4, a majority of prospective elementary teachers lack confidence in their science knowledge, which for many has its origins in their lack of success in science during their schooling, particularly in their feeling that they do not know the right answer. Harlen's (2001) description of "productive questions" in science is a useful tool for formulating science methods course assignments and helping methods students learn about the diversity of question types (see Chapter 7).

The Distribution of Questions. This aspect of questioning is pertinent to instructional strategies used in science methods classes, and important for prospective teachers to be aware of in their own science teaching. Instructors at all levels often call on a few target students, who volunteer readily to answer questions. When asking questions of a whole class, beginning teachers tend to select those students in front of them, and closest to them. Students on the periphery and at the back of the class tend to miss out. Further, boys are often more energetic than girls in bidding for the floor to answer questions, and so tend to be selected more often than girls sitting quietly with hands raised. Changing these patterns requires deliberate changes by beginning teachers in positioning themselves in the classroom, different patterns of scanning the class, and conscious selection of those identified as likely to miss out. Awareness of these tendencies is the first step to overcoming them, but having a colleague observe a lesson and note question distribution is even more helpful. This is especially important, considering teachers' tendency to think that they are, say, favoring girls, only to find from observations that energetic boys are still claiming the greatest share of questions.

An instructional strategy for your methods class that can emphasize this aspect of questioning is to model poor instances of question distribution, and then ask the prospective teachers whom you ignored how they felt. You might also explore ways to get more students involved in answering by asking them to turn to a neighbor and share their answer, or asking them to share ideas in small groups and then calling on individuals randomly from those groups to answer in the large group.

Question Wait Time. The normal classroom transaction pattern when a teacher asks a question is Teacher Question—Student Answer—Teacher Response. The time a teacher waits before selecting a student to answer the question, is called the wait time. Very short wait times of less than a second are common, especially with closed questions. This may well be suitable for recall questions, but does not give students time to reflect if an open question is asked. Rowe (1978) showed how extending the wait time led to a number of educational gains. In school contexts a wait of at least 3 to 5 seconds is desirable, but in a tertiary setting a wait time of up to 15 or 20 seconds can be helpful. This seems an awfully long time, so your prospective teachers need to be aware of what you are doing and why. Such a long wait time is particularly unsettling for the methods student who has learned to opt out of engaging in thinking during class, and prefers to just note the instructor's comment for later memorization.

Gains in learning have also been noted if the transaction pattern is changed to Teacher Question—Student Answer—Student Comment—Student Comment—Teacher Response. This can be achieved if the instructor waits for a while after the first student's answer (called wait time 2). The normal teacher transaction after the student answers a question is to either affirm a correct answer, or to reject an incorrect one. To avoid a direct rejection, many instructors simply choose another person to answer the question—the implicit message is that the first answer was incorrect. Increasing wait time 2 gives other students the opportunity to provide a different answer or to comment on the first one. However, your prospective teachers need to be made aware that you are changing the transaction pattern, and will initially need encouragement by a prompt like, "Anyone else?" or "What do you think of Rose's answer?" Another useful prompt given before asking the question might be, "I would like to hear several people's answer to this question."

In a science methods class, it is essential that you use both types of wait time during instruction. This allows the prospective teachers to experience the benefits of longer wait times for their own learning, and also provides a model for them—provided the instructor's modeled actions are explicitly pointed out and not left implicit, in the hope that the prospective teachers will notice.

Technique for Giving Instructions

Giving instructions is an important aspect of science instruction, and will feature in the prospective teachers' own science teaching, as well as in activity and laboratory work during science methods classes. Instructions are mostly related to safety rules, a laboratory skill, or a procedure for conducting an investigation. Consequently, giving instructions is an important part of a science methods course that needs to be modeled for the prospective teachers. There are two considerations regarding giving instructions: the actual set of instructions, and the reasons for them as perceived by the students.

First, the actual set of instructions, as when giving explanations, needs to be brief, logical, clear, and unambiguous. Giving instructions verbally can be effective, but it is a good idea to supplement this with a written list. Sometimes a written list is more appropriate, such as when the methods students are working in small groups, each group with a different task. Written lists lend themselves to a bullet or numbered list format, which can help the instructor keep the instructions concise. Written instructions are very common for outlining laboratory procedures, but this can create potential problems if the students perceive the task differently from the instructor.

Second, the perceived reasons for instructions can be problematic. Many years ago, Tasker (Tasker & Freyberg, 1985) found that school students' perceptions of the purpose of an activity, and the sequence of steps in doing it, were frequently very different from those of the teacher. This resulted in students "going through the motions" to complete the instructions without learning anything from the activity, or relating it to other work. Being first to finish was a common motivator, even though hands-on activity was preferred over other classroom practices. Little research has been conducted on this since Tasker's studies, but our anecdotal experience suggests that the problem persists both in schools and in science methods courses. For instance, we have seen methods students go into "student mode" when they enter the room, and dutifully follow instructions but not think about what they are doing or why they are doing it. Many are even not aware of the pedagogy the instructor is using unless the instructor specifically draws their attention to it.

Further, Tasker found that merely explaining the task and the instructions did not necessarily resolve the matter. The more successful techniques involved students and teacher cooperatively planning activities, including instructions, from situations meaningful to the students. Therefore, for laboratory and hands-on activity instructions, this technique should involve more than just giving students the set of instructions: the purpose for the activity and the set of instructions should preferably be jointly planned by both the students and the teacher. Sometimes this is not practical, but it does fit better with inquiry-based approaches. Similarly, the science methods instructor needs to explore ways of having prospective teachers engage with both the content and pedagogy of the methods course in meaningful ways. Taking a reflection orientation (see Chapter 3) guides such strategies.

The Direct Teaching Strategy

This strategy is based mainly on the technique of giving explanations, and at first glance may appear to be antithetical to the contemporary push for inquiry (NRC, 1996, 2000). However, there are times even within an inquiry framework, when direct teaching is both appropriate and desirable—especially after a need to know has been established, and the concept or idea being taught cannot readily be accessed through student investigation. This strategy would usually only constitute a part of a lesson and would normally be used in conjunction with one or more other strategies. The strategy is especially useful in science methods classes, and is equivalent to giving a mini-lecture. The strategy has three segments:

Focus. A routine or technique to: focus students' attention onto the topic, provide motivation, and to help them recall personal experience pertinent to the topic. For a science topic this could be another strategy such as the investigation strategy (see below), or a technique such as a teacher demonstration. For a science methods session it could be a brief case scenario (see below), or an advance organizer for the mini-lecture.

Presentation. Information is presented using the teacher explanation technique, preferably including examples, analogies, and visual aids. This would be followed by a teacher-led discussion to help the students think about the information, relate it to their personal experience, and provide the teacher with feedback about the sense they are making of the concept or idea being presented. If the explanation takes more than about 5 minutes, it should be broken into parts with the discussion component interweaved.

Application. The students engage in a task where they apply the information to a real-life context. This could be a minor small group task, or an individual written or online task.

The application of this strategy to both science and science methods classes is self-evident. However, in our experience it has a tendency to be over-used in tertiary methods classes. Ideally, it should constitute just one of a number of strategies used during a methods class, but unfortunately could stretch to become an hour-long lecture. You have probably recognized our belief that lengthy lectures should play only a very minor role in science methods classes, perhaps being used in an introductory class to cover administrative details, or to inspire and motivate students.

Predict–Observe–Explain (POE) Strategy

This is a well-known strategy (e.g., Palmer, 1995) with four main segments, three of which give it its name. The POE strategy would commonly fit within one lesson unless the observe segment was lengthy and spilled over into a second lesson.

Focus. Similar to the focus in the direct teaching strategy above, this introduces the topic and acts as a motivator and attention-getter, but importantly also sets up the situation for the next segment, making predictions. To do so, students must in their minds link to previous experience, or they will guess rather than predict. A short hands-on activity or teacher demonstration would be suitable alternatives depending on the topic.

Predict. In a science lesson, the teacher invites the students to predict what will happen if an aspect of the event is changed. If the event is very well known, the students may be shown the materials and invited to predict what will happen following a particular action. For instance, what would occur if an inflated balloon were pricked with a pin? Predictions should preferably be recorded individually or collectively. It is also advisable to invite the students to give reasons for their predictions. In a science methods session, there are several possibilities. If a suitable science lesson case scenario were used in the Focus segment, the prospective teachers could predict what would happen as a consequence of each of the teacher's possible responses in the scenario.

Observe. The students are then asked to observe as the action is carried out: the aspect of the event is changed, or in the above example, the balloon is pricked. Their observations can be individually recorded, but at least should be shared in the whole group. Observations should be related to the earlier predictions. In the science methods case scenario, the teacher's choice of action would be outlined, and the consequences noted.

Explain. The students are invited to explain what happened and why, with particular attention to the alignment of their predictions with the observations. Explanations can be shared in the whole class, but are best done using a routine such as the "think-pair-share" routine. The science methods case scenario should ideally include an explanation by the teacher for her actions, which the prospective teachers can then discuss.

The POE strategy can be used in methods classes to teach a science topic, so that it serves a dual purpose of teaching some science and modeling a powerful teaching strategy.

The Investigation Strategy

This strategy, or its variants, is a common one used in science lessons. It is also a powerful strategy when used in science methods classes to teach and model both the strategy and a science topic. Although prospective teachers may have taken science in high school or College, it would be unwise to assume that they are competent in any of the segments of this strategy: many would need assistance in, say, planning an investigation, or in data processing. If used to scaffold a science methods session on some educational topic, it works best if the "data collection" segment is done as a field experience in a science classroom. For instance, in a science methods class, this strategy could be used to work through the teaching of a science topic that the prospective teachers will teach in pairs to a small group of elementary students during a field experience. How this might fit with each segment of the strategy is outlined at the end of each segment.

The strategy is mainly hands-on, and historically has been used extensively in elementary science, often called inquiry or discovery.

Unfortunately, as discussed earlier, these terms convey a variety of meanings and refer to a range of different strategies, so we prefer to avoid them to minimize confusion.

The strategy fits comfortably within one lesson, but could be part of a lesson or could be spread across two lessons—especially if data collection was time consuming. The strategy should always be followed by another strategy involving application or further investigation. There are four main segments:

Focus. As in other strategies, the Focus introduces the topic and acts as a motivator, setting the scene for the subsequent investigation. It should also provide opportunity for students to recall relevant personal experience so the investigation can be suitably contextualized. For a science topic, it can take any number of forms, such as a teacher demonstration or a teacher-led discussion based on a picture or short movie clip; but preferably would be a brief exploratory hands-on activity. Students should be encouraged to share their ideas and previous experiences during this segment.

In the science methods teaching example, this Focus segment may, after an explanation of the teaching assignment, involve the prospective teachers finding out elementary students' existing ideas about the science phenomenon/topic that will be taught; and then identifying a few activities that may help address any misconceptions.

The Setting. Taking account of Tasker's findings (Tasker & Freyberg, 1985, see above), this segment is designed to help the students identify the purpose of the next segment, and to clarify the procedure that they will follow. The goal and/or procedure may be drawn from a laboratory manual, activity description, a teacher-posed question, or a question asked by the students themselves (see teaching models below), but needs to be personalized by the students. This is maximized if the students are themselves involved in planning the procedure, under the guidance of the teacher. The following transcript from a grade 4 lesson on "Soils" illustrates this. The purpose of the activity was to see which soil types best retain water, as part of a larger investigation into differences between soils.

TEACHER: OK. We've decided that we will pour some water onto each soil type and catch the water that goes through. We have said that we need to use the same quantities of soil and water for each test. That means we need to put the soil in something to let the water out of the bottom, but the soil can't get out. What could we use?
JOCYLENE: When Mum is cooking she uses a thing like with a bit of fly screen...
RHONDA: A sieve.
CHUCK: We could use a funnel that has a wire thing at the bottom.
TEACHER: Does everyone know what Jocylene and Chuck mean? [She draws each on the board with a brief explanation.] Which would be better to hold the soil and pour water through?

SEVERAL STUDENTS: A funnel.
TEACHER: Well we have some funnels in the school [showing one] so we can use them, but there is no wire in the bottom. How could we stop the soil falling out of the bottom?
ROSS: Put a cork in it.
RHONDA: Then the water wouldn't get out.

It is advisable to compile a written outline of the procedure so students can refresh their memories about what to do while they work. Only use a worksheet if it is developed from the students' plan (which means a time delay between this segment and the next)—a preprepared one signals that the planning exercise had no real purpose and the students' opinions are not valued.

In the science methods example, the prospective teachers would work in their teaching-pairs to plan their teaching assignment, taking into account the likely misconceptions that may emerge. The instructor would monitor their plans and provide feedback.

Data Collection. In a hands-on activity (or series of activities) the students collect data using the procedure decided in the Setting segment. This may entail taking measurements, in which case use of the measuring instrument may need to be taught prior to the activity. For instance, in the above lesson extract, the students were taught how to measure water using a measuring cylinder. Recording of the data may also be an important aspect of the activity, depending on its complexity and the students' ages. The recording format should be included in the planning discussion during the setting segment.

In the science methods example, the class would adjourn and resume in a school setting, with the pairs of prospective teachers each assigned to a small group of students. They would then teach their prepared science lesson.

Data Processing. The students now try to make sense of the data they have collected and perhaps recorded. This may entail transforming the data in order to make it more understandable, such as constructing a table or a graph. Students will usually need assistance doing this. In sense making, they should be relating the data to what they already know and to the goal of the investigation determined in the Focus segment. Students should be encouraged to make inferences and/or hypotheses arising from the data, and which should be related to the science concepts underlying the investigation.

In the science methods example, the prospective teachers would convene after teaching their science lesson to reflect on, and discuss, their teaching experience. Some general principles should be extracted from the experience, if possible.

Some Comments. This teaching strategy is drawn upon frequently in an inquiry-based classroom, so it is essential to make it a specific focus, including being modeled, in a science methods course. Note that prospective teachers would rarely have engaged in the type of planning outlined in the Setting segment during their own schooling, so it is highly desirable for them to experience it during their methods class as science learners, with the features being modeled made explicit. It is also important for your methods students to distinguish this strategy from others, such as the one often called guided discovery. Although this term describes a variety of strategies, it usually refers to a highly structured teacher-led activity where students engage in a hands-on activity, by following the procedure provided by the teacher or a laboratory manual. A common expectation in guided discovery is that the activity will provide the students with "the answer" to the problem they are investigating (i.e., they will "discover" it). Unfortunately, this is often not possible from just doing the activity, so students start guessing, and finally the teacher may resort to telling them the desired answer. Students become adept at recognizing this pattern, and frequently just wait for the teacher to tell them the answer rather than trying to think it through. We suggest that this so-called guided discovery strategy should be used carefully, and only occasionally. Your prospective teachers should be made aware of the differences between so-called guided discovery and the Investigation Strategy, and the pitfalls associated with guided discovery.

The KWL (KWHL) Strategy

This strategy and its variant are named by the acronyms for the main segments: "What do you already *Know*?" "What do you *Want* to learn?" and "What have you *Learned*?" The H in the KWHL variant stands for "*How* are you going to learn it?" (Iwasyk, 1997). The strategy would normally span a lesson, and could sometimes span two lessons. The strategy is widely used in many subjects, including science, and would be suitable in a methods class for a science investigation or an exploration of some aspect of science teaching and learning.

Focus. As in other strategies, an introduction to the topic of study is necessary, but not included in the acronym name. In science, this is preferably a brief hands-on activity or teacher demonstration.

Know. Consistent with constructivist views of learning, students are first asked to identify what they already know about the topic. This can be done individually, or collectively as a class. Brainstorming and concept maps are useful techniques to help students access a comprehensive view of their existing knowledge of the topic. It is best if what they already know is recorded, perhaps in a class concept map. Some visual stimulation such as this can help in the next segment.

Want. The students are invited to state what they would like to know about the topic. This should be done individually, and recorded. The individual queries can be used for personal learning goals and individual investigations. Alternatively, students may then share their queries in order to compile a class short-list that is then used as the basis for investigations. Those queries being addressed should be specifically identified, and where appropriate assigned to small groups or individual students.

How. This is a later variant to the original strategy that reportedly enhances learning. Students are asked to plan how they will obtain the information needed to answer their query. In science, students may need encouragement to plan a hands-on investigation (their first idea tends to be to check a book or the Internet), and help in working out the procedure to follow: sometimes an activity plan can be found in a book or online, but even then students may not readily understand how the activity might relate to their own query (see the comments on planning an activity in the Investigation Strategy).

Learned. In this final segment, students are asked to identify what they have learned. While their evaluation should first be in terms of their original personal or group query, they may well have learned other things that were not part of that original focus, and should be encouraged to consider this as well. It is advisable that they record what they have learned to facilitate comparison with their learning goals.

The Case-Based Strategy

This strategy can be used both in science and society topics, and science methods sessions. It has been used in teacher education for many years, and has more recently gained acceptance in science lessons, mainly those focusing on the history and nature of science. Because it is an especially powerful strategy for prospective teachers to examine particular teaching contexts and understand what is happening, that version of the strategy is outlined here. It is fairly easy to adapt it for a science lesson exploring a particular historical science discovery or controversy.

However, we first need to clarify what we mean by a "case." In this strategy as applied in a science methods class, a case is a brief description of a lesson or lesson segment, usually the latter. It could be a verbal or a written description (real or fictional), or could be a video of an actual classroom. If some or all of the methods class is conducted at a school site, it could be a "live" lesson observed by the prospective teachers. Accompanying the lesson description is usually some contextual information, an explanation or commentary by the teacher or some other expert, and perhaps some focus questions. The nature of the lesson segment is chosen to fit with the topic being addressed, and may focus on, for instance, a particular teaching technique, a teaching strategy, or a critical incident. In essence, the case and accompanying material constitutes a narrative or story. (Resources for written and videocases are included in the

"More to Explore" section of this chapter, and an example of videocase instruction is included in Chapter 3.)

The seven segments are best covered during one session. There are a number of variations to this sequence.

Focus. The instructor introduces the topic around which the case is built.

The Context. The case context is outlined using the accompanying material. This can be done individually, in small groups, or as a large group.

The Case. The case instance is presented to the students. Depending on the medium of presentation and length, the presentation could be paused or even broken into subsets. This allows the instructor to point out important aspects of the case to note, or for different sets of questions to be addressed. If using a video recording, it is often better to play the whole case and then replay it with the pauses.

Focus Questions. The questions for reflection are introduced, ensuring that the students understand what the questions are asking, and to what aspect(s) of the case they refer.

Reflection. The prospective teachers reflect on and discuss the case in small groups, using the questions as a stimulus. Subsequently sharing small group answers/decisions in the larger group can be useful. It is important to encourage the prospective teachers to relate their discussion to both their own practical experience, and theoretical views from their preservice courses.

Commentary. The explanation or commentary accompanying the case is examined. The prospective teachers are invited to compare and contrast their own deliberations with the explanation/commentary. This is best done in the whole group.

Review. The instructor and the prospective teachers summarize the alternative perspectives that emerged from the discussion. Often there is no one "right" answer, but an overall consensus view may emerge.

A very effective variation to this strategy includes having the teacher from the case respond to questions about the case posed by the prospective teachers.

The outline of routines, techniques, and strategies discussed here is but a selection of many possible ones. We have included them because we think that they are the most important to include in a science methods course, but you may wish to add others. These routines, techniques, and teaching strategies tend to be suitable for use within single lessons, or in the case of strategies, possibly spread over two lessons. In the next section, we explore ways to structure a whole unit of work using teaching models.

> **It's Your Turn**
>
> Review the list of instructional strategies you made at the beginning of this chapter.
>
> *Which of the above techniques and strategies were included?*
> *Are any of these techniques and strategies taught in other courses in your institution's preservice program? If not, should they be?*
>
> Devise an action plan so the inclusion of the routines and techniques can be accommodated.
> Revise your list of instructional strategies, adding any techniques and strategies that you now think are important for you to specifically include in your methods class, and to model for your students.

Teaching Models

Teaching models, or teaching approaches as they are also called, consist of a sequence of teaching strategies, techniques, and routines designed to provide a learning scaffold across several lessons; most often for a unit of work focused around a science topic. That is, six to eight lessons would typically be needed for the full sequence of steps in these teaching models. Since they tend to be used for science topics, they should be used in science methods sessions as exemplars to teach prospective students some science, and at the same time to model how the teaching models can be used. There are quite a number of teaching models and approaches that have been suggested in the literature, but few have become widely used. We have therefore restricted our discussion to two of those more widely adopted, but urge readers to explore other models that they feel have merit. Indeed, we would prefer that prospective teachers be encouraged to develop their own teaching models that are most suitable for the teaching situations in which they find themselves. The ones outlined here could be viewed as useful starting points, and are highlighted in Activities that Work 1 and 2.

Teaching models used for specific science topics are instances of a complex mix of aspects of science PCK. Because they draw on a number of routines, techniques, and strategies, they are linked to general pedagogical knowledge. However, the teaching models outlined here tend to be mostly associated with science teaching, and are consequently also associated with PCK for science teaching. Science methods instructors therefore draw on their PCK for science teaching associated with these teaching models in order to develop PCK for teaching science teachers involving instruction about the models and modeling them in practice. A key aspect of this instruction is showing prospective teachers how, in each teaching model, routines, techniques, and strategies are carefully sequenced to form a learning scaffold for school students. Developing a learning scaffold within and across several science lessons is one of the most difficult aspects of PCK development for prospective teachers. Similarly, we have found that many science methods instructors struggle with developing PCK for teaching science teachers related to instruction about teaching models incorporating learning scaffolds.

These teaching models originated from constructivist views of learning. We therefore suggest that any teaching models developed by your prospective teachers should demonstrably be based on constructivist or other learning theories that are justifiable in the circumstances. Prospective teachers often have difficulty specifically relating practice to learning theory, so may benefit from consulting the literature where this has been done (e.g., Appleton, 1993). Note that an important feature of each model is some means of identifying the students' existing ideas relating to the science topic under investigation. While this can be done as a precursor to a unit of work, it is more manageable if it is incorporated into the unit itself. These ideas then inform aspects of the teaching in the remainder of the unit.

Two further comments are necessary. (1) Some authors suggest that these teaching models should be used within a single lesson, i.e., they treat them as strategies (such as those outlined in the previous section). While this is possible, we have found that they are far more powerful when used as teaching models that scaffold learning across several science lessons. (2) A consequent practical matter for dealing with the teaching models in science methods classes is that there is usually insufficient time for the teaching models to be modeled in entirety. This means that they need to be truncated in delivery, but without their pedagogical meaning and relationships to enhancing learning being lost. The models described below are outlined for teaching science topics, as this is how we suggest they be used in a science methods course: teaching a science unit to your prospective teachers while simultaneously providing instruction about the teaching models and modeling them.

The Interactive Approach

This teaching model, also called the "Question-Raising Approach" by Wynne Harlen (1985), had its origins in the Learning in Science (Primary) Project, in New Zealand (Biddulph & Osborne, 1984). The teaching model has been extensively researched (e.g., Chin & Kayalvizhi, 2002; van Zee, Iwasyk, Kurose, Simpson, & Wild, 2001), with a number of publications promoting slight variations of it (e.g., Faire & Cosgrove, 1990; Fleer, Jane, & Hardy, 2007). At first glance, novices (and even experienced teachers) may feel that the unit could go in any direction, and that the teacher would lose control and have management problems. This is possible, but with careful preparation and planning, it is unlikely. In the many hundreds of times we have used this teaching model in the classroom, it has never gone in an unanticipated or unfruitful direction, and we have never lost control of the class. The first of the seven steps in this variation of the model (adapted from Appleton, 1997) emphasizes teacher preparation rather than the usual first step of engaging students.

Preparation

After selection of a science topic for the unit (e.g., Magnetism), an essential component of teacher preparation involves clarification of the teachers' own understandings of the topic, in order to identify any knowledge gaps, and importantly, any misconceptions they may hold. The preparation also entails researching possible student queries (see step 3), and then assembling likely resources for the unit suggested by those queries, including investigative activity ideas and materials.

The Preparation should also involve finding out probable student misconceptions about the topic, which can be a helpful guide for selecting likely activities. Although specific misconceptions held by students in the class could be identified before the unit commences (e.g., by using concept maps or conversations),[5] this is not really necessary since steps 2, 3, and 4 provide insights into these. Note that prospective elementary teachers tend to hold similar misconceptions to those held by 14–15 year olds—even science majors in university.

Exploration

The students are first provided with a brief verbal advance organizer, ideally supplemented by pictures or actual materials. The organizer includes introducing the topic, and engaging in a class conversation to get a feel for what the students already know. For instance, for the topic "Electric Circuits," a battery and a torch bulb would be shown, and students asked to identify each and give a brief explanation of what they do. This also provides an opportunity for the teacher to explain and/or demonstrate any skills needed for the activity. This is then followed by an exploratory, hands-on activity that introduces them to the topic.

The exploratory activity should be chosen with care, as it can influence the types of questions that students raise (see step 3). For example, consider these alternative activities for "Magnetism": (a) using a magnet to sort a variety of objects into those picked up by the magnet and those not, and (b) seeing what happens to a magnetic compass when a bar magnet is

---

5. Conversations are one-to-one or one-to-small group conversations with students in order to probe their ideas. For instance, see Interview About Instances (Osborne & Freyberg, 1985).

brought close. While some questions would be common, activity (a) is likely to generate questions about the nature of materials that are magnetic and why this may be so, whereas activity (b) is likely to generate questions about compasses and how they work.

Students' Questions

Once the exploratory activity has been completed, students are invited to raise any questions about the topic for which they would like an answer. Emphasize that their questions should be ones where they do not know the answers. Questions can be raised using a whole class routine, or a small group one such as the think-pair-share routine. Keep a record of the questions asked, preferably where all students can view them. All questions suggested should be accepted without judgment or comment, unless it is a very trivial question that can and should be answered immediately (e.g., "Why does the wire have an alligator clip on the end?").

When students first engage in this exercise, they sometimes ask questions that do not lend themselves to subsequent investigation (see step 5). Research has shown that after a while, and with appropriate guidance, students learn to ask better questions (Biddulph & Osborne, 1984). Once the list of questions is completed, the teacher and students cooperatively select some questions to be investigated. While it is important that students feel that they have ownership of the questions, it is also important for questions that will promote learning to be selected; so the teacher needs to be skilful in negotiating the choice of questions. The number selected also needs consideration in terms of classroom management. If all students are to work on the same question (perhaps in small groups), then just one (or two if it is a suitable follow-on question) will suffice. If several small groups are to each work on their own questions, then one will need to be chosen for each group. The teacher needs to be aware that the more questions and groups, the greater the management demands. We have seen an experienced and excellent teacher struggle when trying to manage 12 different questions and groups. We suggest that until both students and teacher are familiar with the teaching model, only one question be chosen.

In your science methods class you would be modeling this process, so bear in mind that your prospective teachers will use the number of questions that you (and they) select for investigation as a guide to the number that they should select when teaching. You may be quite adept at managing (say) eight questions being simultaneously investigated by groups of adults, but as neophytes they may struggle managing just two questions being investigated by children.

Students' Possible Answers and Planning Investigations

This next step has two related sections (see the Investigation Strategy above). First, the students are asked what they think might be an answer to the selected question(s). This should initially be done individually, and then shared in the whole group. Reasons for possible answers should also be given and recorded, though accept that some answers may be gut-feeling guesses, just as some students will validly claim that they cannot think of an answer. Second, how an answer to the selected question(s) might be obtained is planned. If a tentative answer is available, the planning should initially focus on how to test if the tentative answer is a good one.

Students will not be able to plan investigations without a lot of help and support, unless they are particularly gifted and/or experienced in this way of working. It is most fruitful if the teacher and students cooperatively plan the investigations—student ownership is important. Many of the possible investigations that the teacher had foreseen during the first step (Preparation) should prove useful, provided that the teacher can point the students to the activities without being too prescriptive. The activities should follow a logical sequence that will provide sufficient information for an answer to the selected question to be arrived at. However, note that activities will not of themselves provide scientific theoretical constructs that might be needed in order to answer some questions, particularly "how come" and "why" questions. Part of the investigation sequence would therefore need to include accessing such theoretical constructs in an understandable form; but this aspect of the investigation should always be toward the end of the investigation sequence: that is, it is best to first establish in the students a "need to know." Procedural details of each activity should also be planned at this stage.

Specific Investigations

Once the planned investigations have been mapped out, and each procedural step identified, the necessary materials can also be identified and assembled. This may take a day or two, so it is wise to plan for such a break after the Planning step. The students should be able to progress through each investigation with normal teacher supervision, though they will need to be reminded regularly of why they are following the procedure, and helped to make sense of their findings in terms of the question(s).

An important role for the teacher during this step is to monitor the students' progress and to provide feedback about how they are going in their quest for an answer to their question. By the completion of the investigative sequence, the students should have successfully decided on what they think is a "good" answer to their question. If the students are still struggling, or have gone down a blind alley, the teacher needs to step in and provide a different learning track that will help them toward the desired goal.

This step may obviously be spread over several lessons, and is the part that would probably need to be truncated in your methods class. However, it is essential that your prospective teachers see you modeling the processes of monitoring progress, providing feedback, and redirecting those who are going down an unproductive learning route. We have found that a video of the instructor using this teaching model (ideally with the same topic and answering the same question) with an elementary class provides an invaluable aid at this juncture: prospective teachers need to see the components of the teaching model being modeled, and have them explicitly pointed out at suitable points during the instructional sequence.

Problem Solving

This is an additional step to those originally proposed by Biddulph and Osborne. We have added it because extensive experience in using this teaching model has shown its value, and it can also be justified theoretically. Once a tentative answer to the selected question has been obtained, it helps student learning if they are able to work further with their newly acquired knowledge (which is not necessarily the answer to their question, but could be knowledge that they acquired along the way to answering their question). The best means of doing this is to provide the students with a new, practical, real-life (to them) problem. An example for the "Electric Circuits" topic would be to ask them to construct a circuit for (say) a doll's house.

Reflection

Students now reflect on the outcomes of their work with the problem (the Problem Solving step), and their tentative answers to the selected questions. At this stage they should be encouraged to make judgments about the quality of their answers and the extent that they align with their initial possible answers (Students' Possible Answers step). The students should communicate their conclusions using a reporting tool, perhaps contrasting their "before" and "after" answers to the questions. This is also a time when the teacher engages in summative assessment, drawing on the student report material, and on conversations held with small groups and individuals.

Note that assessment in the form of written tests is not part of this teaching model, nor should it be. In our experience, the assessment data from the above procedures is far more reliable than written test data.

The 5E Model

This teaching model has also been extensively researched and promoted (e.g., Blank, 2000), having had its origins in the "learning cycle" developed for *Science Curriculum Improvement Study*. It was more recently modified (Bybee, 1997) to this version containing five steps—each beginning with "E"—hence its name. In some educational jurisdictions, the 5E has become synonymous with inquiry, but we feel that this is too simplistic as there are many teaching models that fit the requirements for inquiry in the classroom (NRC, 2000). Some authors have also promoted the 5E as what we have called a teaching strategy, to be used to structure just one science lesson. While this is possible (as it is with all of the teaching models), we do not consider it feasible to adequately deal with all steps if they are each addressed in the one lesson: we believe that the full power of the 5E model resides in its use as a teaching approach to scaffold a whole unit of work—hence its inclusion in this section. The five steps are:

Engage

This step serves a similar purpose to the focus segment in several of the teaching strategies: it introduces the topic and learning task, and captures students' interest. It should also help students make links with the topic and past learning experiences. It could be a hands-on activity (e.g., make a bulb light using a bulb, wire, and battery), a demonstration (e.g., Cartesian diver), a question from a problem scenario (e.g., "We are in a desert wilderness; how can we get drinking water?"), a Concept Cartoon[6] (Keogh & Naylor, 1999), or a puppet-generated scenario[7] (Naylor, Keogh, Maloney, Simon, & Downing, 2007).

Explore

In this step, students engage in a range of hands-on investigations where they "mess around" and become familiar with the materials associated with the topic and concept. This is designed to further encourage links to previous experiences, avenues for students to discuss these with peers and the teacher, and provide a common set of experiences on which to base further concept development. It may also include book or online searching for ideas. The students should be involved in developing investigation ideas, and should be provided with just sufficient help and guidance to guide their learning. By the conclusion of this step, students should be working toward gaining an understanding of the concept(s) underlying the science situation presented in the Engage step, sharing their ideas, and considering alternatives presented by peers.

---

6. A line drawing cartoon of a science situation, with different views about the situation expressed by different children in the cartoon.
7. Hand puppets are used to generate different views about a science situation in the classroom.

Explain

The emphasis in this step is on the students gaining an understanding of the concept(s) underlying the science related to the situation presented in the Engage step. The teacher's role is to help them make sense of what they have discovered in the Explore step, and relate it to established science ideas. This may involve pointing out aspects of the investigations that the students had missed or glossed over, providing information that they had not accessed, making information from other sources understandable in terms of the investigations, and helping students relate this to what they already know. This step should also include opportunity for students to share their understandings.

Elaborate

In this step, it is intended that students work further with their developing understanding of the science concept(s). This may mean working on a problem that can only be solved by drawing on their new knowledge, or it may mean engaging in further investigations where the concept(s) is applied to new situations. Interaction with other students is an important aspect of this step, helping them to construct deeper understandings by having to explain and justify their ideas.

Evaluate

Students now evaluate their own learning of the concept(s): the extent that they understand the ideas, and what they still need to work on. This may involve some form of individual or group reporting, but should include conversations with the teacher where meaningful feedback is provided about learning progress. Summative evaluation can also occur at this time.

Other Teaching Models

As indicated earlier, a variety of teaching models have been proposed in the literature, but have not been discussed here. For instance, another useful model, the OES (Orient–Enhance–Synthesize) model is outlined in ATW 5. We should also mention the Conceptual Change model (Posner, Strike, Hewson, & Gerzog, 1982), as it has been a major focus in the literature. It is based on the notion that students' existing ideas (usually misconceptions) should be replaced by ideas consistent with scientific views. A key step in the model is to deliberately contrast students' misconceptions with the scientific view to create dissonance, and to then provide evidence that the latter view is more plausible, useful and fruitful. This model provided a useful step in science educationists' understanding of learning theory and related pedagogy, but we do not consider that this teaching model continues to have much currency nowadays. For instance, the notion of conceptual replacement is outdated—rather, learners are believed to hold competing ideas in juxtaposition, and invoke the ideas that are triggered by contextual factors: school ideas are invoked by school contexts, and everyday ideas are invoked by everyday contexts. We only deal with this teaching model in our methods course if there is sufficient time, and more as a window into the historical development of other teaching models.

Another historically important, and still useful teaching model not discussed here is the "generative learning model of teaching" suggested by Mark Cosgrove and Roger Osborne (1985) in the follow up work to the *Learning in Science Project*. It follows a similar pattern to other models. The Interactive Approach discussed earlier was a later application to elementary science that emerged from this model.

> **It's Your Turn**
>
> *To what extent are you familiar with the teaching models outlined in this section?*
> *What other teaching models are you familiar with?*
> *How could you model these or other teaching models in your science methods course?*

Teaching How to Plan

A common assumption we instructors make is that planning a unit of work or a lesson is intuitive, straightforward, and easy for our prospective teachers. While this may be true for some, we have found most do not know how to plan science in a way that will enhance student learning. We have therefore found it necessary to include specific instruction on planning. The most effective instructional strategy for doing this is to model the planning process, with the prospective teachers working with us cooperatively so that we jointly plan a unit of work on a science topic for a specific grade, and then plan one or more lessons from the unit.

In particular, we have found that few prospective teachers are aware of how to plan for enhanced learning. We explicitly teach, via modeling and cooperative engagement, the following sequence (see also Chapter 5):

1. Identify for the selected science topic what the students are expected to learn (i.e., the learning goals). This needs to include specific statements of science ideas or concepts (drawn from the official curriculum) appropriate to the grade level. Vague statements are not permissible: e.g., "Students will know about the phases of the moon" is vague, whereas "Students will explain that the phases of the moon are caused

by viewing different proportions of the lighted and dark portions of the moon's surface" is explicit. Further, the statement contains a verb linked to assessment. Of course that exact statement is not what will be assessed, but the idea embedded in the statement will be. Prospective elementary teachers initially have great difficulty doing this, often because they are uncertain about the science ideas themselves. We ask students to examine state and national curriculum standards and to align their chosen learning goals with the standards.

2. Identify what the students already know about the topic, especially any misconceptions. This can be done through past experience with this grade level, investigating research reports of students' ideas, a brief investigation with students to identify their ideas, or the initial focus/exploratory session at the beginning of the unit. We have found it best for our prospective teachers to work with children individually or in a small group to identify their ideas (using techniques such as concept mapping or conversations).

3. Devise tentatively ways of assessing the learning goals—that is, the assessment tools and techniques that might be used. These will be later reviewed in the light of the completed instructional plan.

4. Devise a learning and instructional sequence that will scaffold students' learning, taking the students from what they already know to the desired learning goals. This should be mapped in terms of a series of activities that might include hands-on activities, book and/or Internet research, discussion, asking an expert, and recording/transforming data. The teaching models examined earlier provide a useful guide. These activities can later be organized into specific lessons. Prospective teachers find it very difficult to work out the best way to sequence these activities so that they provide a learning scaffold—they tend to think in terms of activities and lessons, and do not give careful consideration to sequence. For instance, a common notion is to explain a science idea then provide hands-on investigations to "prove" it, whereas we would suggest that students should engage in hands-on activities to explore the idea and then try to reach their own explanation (e.g., see the Interactive Approach) before they are given an explanation.

5. Devise techniques to enable ongoing monitoring of students' learning and progress toward the learning goals—formative assessment (see Chapter 7). Identify where such formal assessment may also be used in summative assessment. Review the summative assessment techniques to ensure that they will provide sufficient information about students' learning without over-assessment.

Conclusion

The instructional strategies outlined in this chapter have been organized around an increasing complexity of teaching actions, from routines, to teaching models. They are mostly drawn from general pedagogical knowledge, and can be transformed into your PCK for teaching science teachers, as well as your and your students' PCK for elementary science teaching. A number of instructional strategies specifically applicable to science methods classes have been embedded in the discussion. These are:

- Teach instructional strategies to prospective teachers by explaining and modeling them.
- Teach science content in the methods course through the modeled instructional strategies.
- Model how to plan a unit and a lesson using examples where you and your prospective teachers cooperatively and jointly plan.
- Make all modeled teaching behavior explicit.
- Incorporate brief explanations of the theoretical bases for aspects of the instructional strategies modeled (e.g., learning theory).
- Use case narratives (including video) when possible to support and exemplify your teaching about the instructional strategies.
- Have prospective teachers use some of the modeled science instructional strategies in science teaching situations—preferably with children in a field placement where ideally they teach in pairs to a small group of 4–12 students.[8] Ensure that you check their intended teaching plans before they commence.

> Sam, Joan, and Noeline met partway through their teaching of the elementary science methods course. "Let's share one idea that worked in our teaching recently," suggested Noeline. Joan jumped right in. "I wanted my students to learn the 5E model, so I decided to engage them in a 5E session themselves. It was brilliant! After the session, I helped them describe what the teacher and students did at each phase. They seemed to understand. Of course the real proof will be when they try to design their own 5E lessons." Sam nodded, "I have found that asking my students to think like science learners and then like science teachers is a powerful strategy indeed." Noeline realized that the instructional team was becoming more cohesive due to their shared reliance on a set of instructional strategies and approaches.

8. Working in pairs with a small group is a deliberate strategy to reduce and minimize anxiety about teaching science, and to maximize their chances of a successful teaching experience.

More to Explore

Goodrum, D. (2004). Teaching strategies for science classrooms. In G. Venville & V. Dawson (Eds.), *The art of teaching science* (pp. 54–72). Crows Nest, Australia: Allen & Unwin.

Hackling, M. (2004). Investigating in science. In G. Venville & V. Dawson (Eds.), *The art of teaching science* (pp. 88–104). Crows Nest, Australia: Allen & Unwin.

Harlen, W., & Qualter, A. (2004). *The teaching of science in primary schools.* London: David Fulton.

Louden, W., & Wallace, J. (Eds.). (2001). *Dilemmas of science teaching: Perspectives on problems of practices.* London: Routledge.

Tippins, D. J., Koballa, T. R. Jr., & Payne, B. D. (2002). *Learning from cases: Unraveling the complexities of elementary science teaching.* Boston: Allyn & Bacon.

Examples of Videocases

Annenberg Media (http://www.learner.org/) has a collection of videocases of science classrooms at various levels that can be useful for the elementary science methods course.

Gallery of Teaching and Learning (http://gallery.carnegiefoundation.org/) provides digital representations of knowledge related to teaching and learning crafted by participants in the Carnegie Foundation's programs for teaching.

Reflecting on Elementary Science (http://roes.missouri.edu/) is a space for prospective elementary teachers to observe and reflect about the teaching of elementary school science for conceptual change through three interactive videocases.

References

American Association for the Advancement of Science. (1993). *Benchmarks for science literacy.* New York: Oxford University Press.

Appleton, K. (1993). Using theory to guide practice: Teaching science from a constructivist perspective. *School Science and Mathematics, 93,* 269–274.

Appleton, K. (1997). *Teaching science: Exploring the issues.* Rockhampton, Australia: Central Queensland University Press.

Bell, B. (2007). Classroom assessment of science learning. In S. K. Abell & N. G. Lederman (Eds.), *Handbook of research on science education* (pp. 965–1006). Mahwah, NJ: Lawrence Erlbaum.

Biddulph, F., & Osborne, R. (1984). *Making sense of our world.* Hamilton, New Zealand: Waikato University, Science Education Research Unit.

Blank, L. M. (2000). A metacognitive learning cycle: A better warranty for student understanding? *Science Education, 84,* 486–506.

Bybee, R. (1997). *Achieving scientific literacy: From purposes to practices.* Portsmouth, NH: Heinemann.

Chin, C., & Kayalvizhi, G. (2002). Posing problems for open investigations: What questions do pupils ask? *Research in Science and Technological Education, 20,* 269–287.

Cosgrove, M., & Osborne, R. (1985). Lesson frameworks for changing children's ideas. In R. Osborne & P. Freyberg (Eds.), *Learning in science: Implications of children's science* (pp. 101–111). Auckland, New Zealand: Heinemann.
Faire, J., & Cosgrove, M. (1990). *Teaching primary science*. Hamilton, New Zealand: Waikato Education Centre.
Fleer, M., Jane, B., & Hardy, T. (2007). *Science for children: Developing a personal approach to teaching* (3rd ed.). Sydney, Australia: Pearson Education.
Harlen, W. (1985). *Teaching and learning primary science*. London: Harper Row.
Harlen, W. (2001). *Primary science: Taking the plunge* (2nd ed.). Portsmouth, NH: Heinemann.
Iwasyk, M. (1997). Kids questioning kids: "Experts" sharing. *Science and Children, 35*(1), 42–46, 80.
Keogh, B., & Naylor, S. (1999). Concept cartoons, teaching and learning in science: An evaluation. *International Journal of Science Education, 21*, 431–446.
National Research Council. (1996). *National science education standards*. Washington, DC: National Academy Press.
National Research Council. (2000). *Inquiry and the national science education standards*. Washington, DC: National Academy Press.
Naylor, S., Keogh, B., Maloney, J., Simon, S., & Downing, B. (2007, July). "We talk more, we listen more, we learn more": Changing the culture in primary science classrooms. Paper presented at the annual conference of the Australasian Science Education Research Association, Fremantle, Western Australia.
Nuthall, G. (1999). The way students learn: Acquiring knowledge from an integrated science and social studies unit. *Elementary School Journal, 99*, 303–341.
Nuthall, G. (2001). Understanding how classroom experience shapes students' minds. *Unterrichts Wissenschaft, 29*, 224–267.
Osborne, R., & Freyberg, P. (Eds.). (1985). *Learning in science: The implications of children's science*. Portsmouth, NH: Heinemann.
Palmer, D. (1995). The POE in the primary school: An evaluation. *Research in Science Education, 25*, 323–332.
Posner, G. J., Strike, K. A., Hewson, P. W., & Gerzog, W. A. (1982). Accommodation of a scientific conception: Toward a theory of conceptual change. *Science Education, 66*, 211–227.
Rowe, M. (1978). *Teaching science as continuous inquiry: A basic instructor's manual* (2nd ed.). New York: Holt, Rinehart & Winston.
Smith, L., & Land, M. (1981). Low-inference verbal behaviors related to teacher clarity. *Journal of Classroom Interaction, 17*(1), 37–42.
Tasker, R., & Freyberg, P. (1985). Facing the mismatches in the classroom. In R. Osborne & P. Freyberg (Eds.), *Learning in science: The implications of children's science* (pp. 66–80). Auckland, New Zealand: Heinemann.
van Zee, E. H., Iwasyk, M., Kurose, A., Simpson, D., & Wild, J. (2001). Student and teacher questioning during conversations about science. *Journal of Research in Science Teaching, 38*, 159–190.

Chapter 7

Assessment Strategies for the Elementary Methods Course

> When I saw my colleague Rhonda in the hallway, she was intent on getting to the work room. "I've got to get my final exam copied before class," she said breathlessly. As I followed her to the copy machine, she asked, "When are you giving your final?" "I'm not, actually," I admitted. She looked surprised. "How will you know what grades to give?" I explained that I had used some assignments throughout the semester to assess what students were thinking and other assignments in which they applied their understanding to hypothetical and real classroom problems. "This week," I related, "they are coming to my office for their exit interviews and they'll describe their progress in the course using their portfolios." Rhonda paused, intrigued by what must have seemed a radically different way of assessing methods students. "Let's talk more about this after the semester ends," she suggested.

Although testing is a common strategy for assessing student learning in college courses, tests often fail to capture the type of "thinking like a science teacher" that we expect of our methods students. Tests may provide a convenient method of establishing a grade, but they typically do not require students of science teaching to apply their learning to problems of practice. Thus assessing students in the science methods course requires forms of assessment different from those used in other college courses, such as science content courses.

The purpose of this chapter is to help you build your knowledge of assessment as applied to the elementary science methods course setting. Thus you will develop PCK for assessing science teachers responsive to your orientation (see Chapter 3). In this chapter, we examine the purposes and principles of assessment that apply to the elementary science methods course, and describe a variety of strategies for assessing elementary education students. Specifically we will address diagnostic, formative, and summative assessment strategies, share sample assignments, and discuss scoring and grading.

Purposes and Examples of Assessment in the Elementary Science Methods Course

Although one obvious purpose of assessment in any course is to derive final grades for students, assessment can serve a number of other purposes as well. In particular, assessment in the methods course:

- helps instructors understand students' incoming ideas;
- helps instructors gauge student progress toward achieving course goals;
- provides data on which instructors can base instructional decisions;
- assists students in developing and applying knowledge, and practicing skills;
- informs students about what they know and do not know so they can take action on their learning.

These purposes lead to three main types of assessment:

- *Diagnostic Assessment.* Diagnostic assessment occurs at the beginning of the course or unit of study. These assessments provide data about students' existing knowledge and beliefs about teaching and learning science in the elementary classroom, and provide guidance for instruction (see Chapter 4; ATW 4 and 8).
- *Formative Assessment.* Formative assessment occurs during instruction. Assessments embedded in instruction help students learn new material and provide assessment data to instructors and to students. This feedback helps instructors think about the kinds of instructional interventions that need to occur. This feedback helps students monitor their own learning (see ATW 8).
- *Summative Assessment.* Summative assessment provides documentation of student learning at particular points in time, most typically at the end of a unit of study or the end of the course. Summative assessment information is often the basis of final course grades.

All three types of assessment can help students think about their own thinking (Abell, 2009), being metacognitive about what they know and do not know and about the progress they are making. Also, all three types of assessment provide feedback to the instructor that can guide next steps in instruction. Furthermore, one assessment may serve multiple purposes. For example, a series of formative assessments may help students perform a summative assessment task. However, for a complete picture of students' learning, all three types of assessment are critical.

We have developed, adapted, and used a variety of assessment techniques and tasks to address these three purposes. In the following sections, we describe some examples of assessment techniques and tasks for each purpose.

Diagnostic Assessment Techniques

Diagnostic assessment can be used at the beginning of the elementary science methods course to document in the incoming knowledge and beliefs of your students related to science content, the nature of science, orientations to science teaching and various aspects of PCK. We have developed and adapted a number of diagnostic assessments for the elementary science methods course. Table 7.1 describes some of these diagnostic techniques, linked to the course content being assessed and to resources in the literature where the technique is described in more detail. A few of these techniques are also described in Chapters 3 and 4. As you will note, several of the techniques can be used to assess more than one area of course content.

We know that students come to our methods classes with many ideas about science teaching developed over many years in the apprenticeship of observation (Lortie, 1975) as students. Oftentimes their ideas are limited in scope or at odds with the views of the methods course instructor. Understanding what students know and believe at the beginning of the course or course module helps the science methods instructor think about ways to challenge students' thinking and broaden their views.

It's Your Turn

Choose one of the diagnostic assessment techniques in Table 7.1 to use in your methods course to find out your students' ideas about science content, the nature of science, orientations to science teaching, or PCK of learners, curriculum, instruction, or assessment. Ask your students to complete the diagnostic instrument and analyze the findings. *What do your students know and believe? How does that compare with your course goals? How does this information inform your instruction?*

Formative Assessment Techniques

Formative assessments are informal ways of gauging student progress toward achieving course goals. They often do not take much time for students to complete, but provide a window into what they understand and where they are having difficulties. Formative assessments can also be

Table 7.1 Diagnostic Assessment for the Elementary Science Methods Course

Methods course content area assessed	Techniques/references
Science subject matter knowledge	*Formative assessment probes.* Students answer various questions related to their understanding of science concepts (see for example, Keeley, 2008). *Concept map.* Students think about the major science concepts within a theme and create a map that links the concepts (Novak, 1998; Novak & Gowin, 1984). *Science autobiography.* Students write about their experiences learning and teaching science prior to the science methods course (Koch, 1990). *Beliefs questionnaire.* Students respond to questions about their ideas related to science concepts (see, for example, science beliefs quiz https://www2.oakland.edu/secure/sbquiz). *Science test.* Students respond to questions about their scientific reasoning in a particular topic area (e.g., Force Concept Inventory, FCI) (for the FCI and other subject specific tests, see http://modeling.asu.edu/R&E/Research.html).
Nature of science knowledge	*NOS card sort.* Students are presented with a set of cards with words or phrases related to a particular theme, and asked to group the cards and explain their groupings (Cobern & Loving, 1998). *Draw a Scientist Test (DAST).* Students draw what they think a scientist looks like (Chambers, 1983; Finson, Beaver, & Cramond, 1995). *Views of Nature of Science (VNOS) questionnaire.* Students respond to questions about various aspects of the NOS (Lederman, Abd-El-Khalick, Bell, & Schwartz, 2002).
Orientations to science teaching	*Orientations card sort.* Card sort task to uncover orientations to science teaching (Friedrichsen & Dana, 2003). *Draw a Science Teacher Test (DASTT-C).* Students draw a picture of themselves as a science teacher at work (Thomas, Pederson, & Finson, 2001). *Science autobiography.* Students write about their experiences learning and teaching science prior to the science methods course (Koch, 1990). *Beliefs questionnaire.* Students respond to questions about their ideas related to science teaching (Hubbard & Abell, 2005).
Aspects of PCK for teaching elementary science	*Lesson preparation method* (van der Valk & Broekman, 1999). Students prepare a lesson or series of lessons related to teaching a particular science topic at a particular grade level on the first day of the course, and revisit their developing PCK as the semester progresses. *Concept map.* Students think about the major pedagogical concepts within a theme and create a map that links the concepts. *Lesson sequence card sort.* Students put lesson activities in order according to a learning cycle teaching approach (Hanuscin & Lee, 2008).

designed to help students be metacognitive about their learning. Examples of formative assessment techniques for the elementary science methods course include:

- *Reading reaction sheets* (Abell, 1992). Before, during, and after reading probes that require students to think and write about what they are reading, and compare what they are reading to their current ideas.
- *Science teaching and learning notebooks/journals* (Abell, George, & Martini, 2002—see ATW 8). In a science teaching and learning notebook, students keep track of their evolving science understanding about the topic under study and they reflect on ways to help elementary students learn the same ideas. The instructor does "spot checks" of journals over the semester, probing the students for deeper thinking.
- *Minute paper/muddiest idea* (Angelo & Cross, 1993). At the end of a given class period, students write down one new idea they learned and one question that remains, or one point that is still unclear (muddy) in their minds.
- *Whiteboards/chart paper group thinks*. When students work in class to discuss their science and pedagogical ideas, they use whiteboards or chart paper to summarize their group thinking and present to their classmates.
- *Accomplishment papers*. Periodically throughout the semester students write a short (one to two page) paper in which they describe a few "aha" moments they have experienced in the methods course and associated field experience.
- *Online "Thinking like a Teacher" reflections*. As part of the course website, students are connected through online discussion boards where they respond to instructor prompts that ask them to compare and make sense of course readings, discussions, activities, and field experiences (see Chapter 8).
- *Peer review*. Certain writing assignments (for example, developing a teaching philosophy) benefit from the peer review process (Bean, 1996) where students in the class read and provide written feedback to the writer.
- *Self-assessment*. Sometimes it is appropriate for students to assess their performance in the class. For example, we ask students to assess their contributions to small group work and to whole class discussions.
- *Submission of draft work*. Students submit drafts of assignments such as potential inclusions in a portfolio (see below), or lesson and unit plans to the instructor, who provides oral or written feedback.
- *Small group conferences and summaries*. The instructor conferences with small groups of students about an aspect of the course (e.g.,

applying constructivist ideas in science teaching, a unit plan). Each student and the instructor then write a brief summary of key points raised during the conference, and exchange summaries.

Formative assessments occur in various ways at many points throughout the course. Typically they are not graded. Grading often defeats a primary purpose of formative assessment—to help students develop their knowledge and think about their progress in learning. Grading can make students feel uncomfortable taking the risk of sense-making in the class, because the stakes are high. Instead of scoring or grading this type of formative assessment, students will benefit from informal feedback in the form of instructor questions and comments. Such feedback provides students with information about how they are doing in the course and gives them opportunities to think more deeply about course topics.

> **It's Your Turn**
>
> Try your hand at formative assessment in your class. For example, use a reading reaction sheet for 2 weeks and see what you learn about your students. Read their reading reactions before the next class. *What do they find surprising in the readings? What are they confused about? How does what you learn from the reading reactions affect what you do in class?*

Summative Assessment Techniques

Summative assessment techniques are typically the major written assignments in the course that demonstrate student achievement of course goals. They often require students to synthesize a number of ideas and apply them to a science teaching task. For example:

- *Content Representation (CoRe) paper* (Loughran, Berry, & Mulhall, 2006). Students decide on the big ideas they will teach in a unit of instruction, and then complete a chart in which they analyze their own content understanding, potential student misconceptions, instructional ideas, and assessment strategies for teaching the ideas.
- *Lesson or unit planning.* Students develop a lesson to introduce a new topic, a demonstration lesson, a lesson that integrates science and language arts, or a unit of work based on a teaching model such as the learning cycle (see Chapter 6 and ATW 1).
- *Curriculum materials evaluation.* Students use a curriculum analysis tool (e.g., American Association for the Advancement of Science, http://www.project2061.org/publications/textbook/mgsci/summary/criteria.htm; or National Research Council, 1999) to evaluate a

module from a published curriculum, or to compare modules from different curricula.
- *Resource file development.* Students compile a set of resources from the Internet, curriculum materials, tradebooks, etc., that will help them teach a particular science topic, and provide a rationale for their selections.
- *Methods course portfolio.* Students create an edited collection of evidence and reflections representing their progress toward achieving course goals (Moseley, 2000). Students participate in an exit interview with the course instructor in which they demonstrate progress in the course by citing evidence from their artifacts.
- *Tests.* These can be used to supplement data about student learning collected from other assessment techniques, but should not be relied upon solely. We believe that constructed response and essay items provide more information about student learning.
- *Teaching philosophy statement* (some call this a "research-based rationale"). A teaching philosophy addresses such things as one's beliefs about science content, students, learning, and teaching. The teaching philosophy can stand alone (which is often applicable to a later job application), or be related to a lesson or unit plan (i.e., why the pedagogy for a lesson or unit plan was selected).

Summative assessments involve greater commitment from the students to generate, and more time for the instructor to evaluate. Typically the instructor scores the final product and assigns a grade. We discuss this process in greater depth later in this chapter.

Principles of Effective Assessment

Not all forms of assessment can address each assessment purpose, but taken as a whole, assessments in the methods course can serve all of these purposes. Yet not all assessment tasks are equally valuable as sources of information about student learning. In order to be most effective, assessment in the methods course should be grounded in principles of effective assessment (Banta, Lund, Black, & Oblander, 1996). We have conceptualized five such principles for effective assessment in the elementary science methods course.

1. Assessment must be linked to course goals.
2. Assessment should provide opportunities for further learning (i.e., be embedded).
3. Assessment should be linked to the jobs of a science teacher (i.e., be authentic).
4. Multiple types of assessments that target various levels of thinking skills at different points in time provide a more complete view of student learning.

5. Assessment should help future teachers be metacognitive about their knowledge and skills in order to self-regulate their learning.

In the following sections, we describe each principle and provide examples from our methods courses that illustrate that principle. We also provide suggestions for you to try out for your elementary science methods course.

Principle 1. Assessment Must be Linked to Course Goals

A major tenet of generating curriculum and instruction is "backwards design" (Wiggins & McTighe, 2005). The idea is that curriculum design begins with learning goals—instructors plan instruction by starting with the endpoint in mind. This same notion applies to designing the methods course. Science teacher educators begin by considering what they want their students to know and be able to do relative to elementary science teaching when they leave the methods course.

Backwards design requires that methods instructors think about three stages of course planning. In the first stage, instructors identify the desired results from students who take their course. They ask themselves what knowledge about science teaching and learning is worth understanding and what skills do science teachers need to develop? This is often a dialectical process in which the instructor brainstorms learning goals and then culls them in order to get to the big ideas, the essential understandings about elementary science they want students to achieve (see Chapter 5, where we talk about the goals for and content of the methods course). In the second stage of planning, instructors determine the kinds of evidence that will be used to establish that students have achieved these desired results. At this point, instructors create summative course assignments, performance tasks, and tests that are aligned with the desired results. These summative assessments provide evidence of student learning in the course. It is natural that some assignments will link to a number of course goals, especially assignments that occur toward the end of the course and ask students to synthesize a number of course ideas. In the third stage of planning, instructors determine the learning experiences and sequence of instruction that will promote student learning (see Chapter 6). These learning experiences include embedded formative assessments that both help students learn and inform instructors of student progress and difficulties in the course.

Table 7.2 demonstrates the backwards design process in action in an elementary science methods course. In this case, the instructor has determined than an important learning goal in the class is that future elementary teachers of science understand how children learn science, including knowing something about the common misconceptions students have in particular concept areas. The instructor has designed a summative performance task, which involves analyzing interviews with children, in order

Table 7.2 Backwards Design in the Elementary Science Methods Course

1. Course goal (What is worth understanding?)	2. Evidence—summative assessment (How will students demonstrate learning?)	3. Learning experiences/formative assessments (How will students build their understandings?)
Methods students will develop a deeper understanding of how elementary students learn science, including what ideas they come to science class with and how those ideas compare to the research on student misconceptions.	Interview assignment—Methods students write up their findings about children's science ideas of a particular science concept after conducting and analyzing interviews about instances or events (Osborne & Freyberg, 1985). They compare their findings with the misconceptions literature and apply their findings to the development of an instructional sequence.	• In class activities (e.g., viewing and analyzing video interviews with children about their science ideas) • Science methods notebook entry (e.g., think about a time you had a science misconception, what did you think and why?) • Online reflection about the field experience (e.g., follow one student through a science lesson. What understandings did he/she exhibit? How does this compare to what we read in class?) • Reading reaction paper (e.g., reading and responding to a journal article about students' misconceptions, such as Lee & Hanuscin, 2008)

to provide evidence that methods students have built PCK about science learners. The instructor plans several in-class activities that include formative assessments to help methods students build this knowledge prior to the summative performance task.

This three-stage sequence is an idealized version of the course and assessment design process. In practice, instructors plan back and forth across these steps as they design the course and major units of instruction. The key aspect of assessment that this planning process conveys is that assessment must be linked to learning goals.

> **It's Your Turn**
>
> Select one of the learning goals that you developed in Chapter 6. *What kinds of evidence would you need to demonstrate that your elementary science methods students had learned this?*

Principle 2. Assessment Should Provide Opportunities For Further Learning (i.e., Be Embedded)

Assessment is often indistinguishable from instruction—it is seamless (Abell & Volkmann, 2006). Seamless assessment is embedded in teaching. It is tied to instructional goals. It is a purposely planned component of the instructional sequence. That is, good assessment activities can also provide the context for further student learning. In the elementary science classroom, assessment opportunities appear while teachers teach. For example, while students work in small groups to carry out a science activity on sinking and floating, the teacher observes their interactions with the materials and with each other. She may ask the small group specific questions that help her to gauge their understanding. She may end the science lesson by asking students to write in their science notebooks how they could explain to their friend why some things sink and other things float. Each of these examples demonstrates assessment that is embedded in instruction. These assessments not only help the science teacher determine what students understand, they also provide opportunities for the students to build their understanding.

This same principle can apply to assessment in the elementary science methods course. To help prospective teachers build their knowledge of science teaching and learning, we find ways to embed assessment in instruction. Imagine an elementary science methods class working to understand how to help children learn science through reading. The science methods course instructor assigns students to read an article from an elementary science teachers' journal, like *Science and Children*, about reading in science (e.g., Abell, 2007), and asks students to complete a reading reaction sheet (Abell, 1992). When they get to class, the instructor gives them each a children's book where sinking or floating is part of the story (for example, *Who Sank the Boat?* by Pamela Allen, 1982), and asks them to apply the suggestions from the article to designing a lesson about sinking and floating using the book as part of the lesson. The instructor circulates and listens to the conversations while the methods students work. She notices them referring to their reading reaction sheets and to the article as they work on their plans. When they share their plans on whiteboards later in class, the instructor notes that each group was able to apply some of the ideas from the article, but no group had used strategies learned in their reading methods course (taken the previous semester) to assist students with the science class reading. The instructor uses this as an opportunity to help the methods students connect what they had learned in previous courses to the elementary science classroom setting.

In this example from the elementary science methods course, it is difficult to distinguish the instructional strategies from the formative assessment strategies. The task of demonstrating knowledge through the lesson design task is an opportunity for students to clarify what they learned from the reading and build new understandings through the application task. While students are building their understandings, the instructor is learning about their learning, and figuring out where to go next in her instruction.

Principle 3. Assessment Should be Linked to the Jobs of a Science Teacher (i.e., Be Authentic)

On a daily basis, science teachers engage in a myriad of activities. They design instruction, manage classroom behaviors, ask questions, develop assessment tasks, interact with parents and administrators, gather resources for instruction, reflect on their own teaching, and so on. They are evaluated on the basis of how well they perform these activities; they are *not* tested. Tests are not authentic tasks for professionals. It is our belief that, if you want to prepare future teachers to think and act effectively, then assessment should be aligned with the kinds of things that real teachers do. In other words, assessment in the methods course should be authentic.

Thus in our methods courses we attempt to design the major course assignment to be authentic in nature, representing the work our students will do some day as teachers. Here are a few examples:

- Design a lesson to introduce the idea that light reflects off objects in certain predictable ways.
- Analyze a unit about erosion from a particular set of curriculum materials. In what ways does it support teachers in teaching erosion concepts? What pieces need to be added or revised to reflect what you know about best practices in elementary science instruction?

- Develop a resource file for teaching about ecosystems.
- Teach a lesson in your field placement that helps students compare two different explanations for a concept (for example, "All sound produced by vibrating objects," vs. "Only some sounds are produced by vibrating objects" (cf. Abell, Anderson, & Chezem, 2000). Write a two-page reflection about your analysis of student learning.

We share other examples of authentic assessment in the elementary methods course in more depth in the "Activities that Work" section of this book.

> **It's Your Turn**
>
> Return to your learning goals and select one. Design a summative authentic assessment that approximates the real work of teachers that will demonstrate the degree to which your students have met this learning goal. Also, think about in-class activities that can serve as formative embedded assessments to help prepare students for this authentic assessment task.

Principle 4. Multiple Types of Assessments that Target Various Levels of Thinking Skills at Different Points in Time Provide a more Complete View of Student Learning

In many college science courses, student assessment consists of exams administered after set sections of the course. These exams provide a window into what students know at that point in time. These exams rely mainly on lower level reasoning (for example, recognizing, recalling, comparing, and summarizing (see Anderson et al., 2001) the content). Cognitive scientists tell us that experts in a field not only remember facts and concepts, they are also able to use those ideas to solve problems. That is, they can transfer those ideas to problem solving in new settings (Bransford, Brown, & Cocking, 1999). The transfer process involves higher order thinking where students analyze, evaluate, and create new ideas in the process of problem solving (Anderson et al.). Thus, if an instructor uses only tests of low-level knowledge, he will get an incomplete picture of student learning.

In the elementary science methods course, instructors attempt to move future teachers toward becoming more expert in their thinking as a teacher. We want them to demonstrate not only that they understand something about science teaching and learning, but also that they can transfer their methods course learning to the elementary science classroom, thus demonstrating that they are developing PCK for elementary science teaching. Assessments aligned with this overarching goal require students to demonstrate higher order thinking processes throughout the methods course. Incorporating a variety of types of assessments at different points in the course allows students to exhibit both their developing comprehension and their ability to apply their understandings to a classroom setting.

Developing the course syllabus (see Chapter 5) is a good place to begin thinking about the types and timing of assessments in your course. Certain kinds of assessments work well early in the course, because they set the stage for your students to develop important ideas about student learning. For example, asking students to develop an Interview about Instances (Osborne & Freyberg, 1985) and use it to interview a few children can help your students recognize that learners often come to science class with misconceptions that can affect their learning. Or methods students can use readily available formative assessment probes (Keeley, 2008) to find out what students already think about a topic that their mentor teachers are going to teach during the field experience. Learning about student science learning through these assessment tasks can form the basis for learning various instructional strategies later in the course. Other course assignments might work best if developed gradually throughout the course. For example, asking students to design a series of lessons that incorporate specific teaching and assessment strategies might be undertaken effectively as a series of small steps with built-in instructor feedback along the way. Still other assignments might work best at the end of the course, as a way for students to synthesize their learning in the course. We have used a few different assignments for this purpose:

- Analyze an elementary science curriculum that a local school district is considering adopting. Write a position paper on which curriculum would be better for the elementary science instruction in your district (see "Tools").
- Create a portfolio of your learning in this course by selecting several work samples that are most meaningful to you, and explain how these work samples demonstrate your progress toward achieving the course goals (see "Tools").
- In a small group, devise a unit of work on a science topic that is based on students' prior knowledge, states explicitly what students are expected to know and be able to do at the end of the unit, and uses an inquiry-based teaching model. Individually write an explanation of why your team selected particular learning experiences and pedagogies. After implementing the unit of work in a field placement (not assessed), write an evaluation of the implementation that includes comments on students' learning (with evidence), and on your own teaching.

Principle 5. Assessment Should Help Future Teachers be Metacognitive about their Knowledge and Skills in Order to Self-Regulate their Learning

As discussed in Principle 2, assessment activities can help science methods students learn new ideas about science teaching and learning. Yet experts, in addition to having more knowledge, are also better at understanding which things they know and which things they do not know. That is, they are able to be metacognitive about their expertise, which helps them guide themselves toward new understandings (Bransford, Brown, & Cocking, 1999). In other words, students can regulate their own learning (Abell, 2009). Self-regulated learners set learning goals for themselves and monitor their progress in achieving those goals. When our methods students are metacognitive, they learn with understanding and retain what they learn.

Learning to be a teacher is a lifelong venture. The methods course is only the beginning. Thus it is important that we equip future teachers with the metacognitive abilities to know when they do not understand something as well as they would like, and the practical skills to seek out resources for their own professional development. One way to do this is to include assessment tasks in your course that ask your students to take stock of where they have been and what they are learning.

Metacognitive questions can be part of diagnostic, formative, and summative assessments. For example, after doing a diagnostic card sort about their incoming science teaching orientations (Friedrichsen & Dana, 2003), an instructor might ask students, "Which ideas do you believe in most strongly? Which ideas are you a bit unsure about?" The same card sort task can be used later in the semester as a formative assessment, where instructors ask students to recognize which of their ideas have changed since the beginning of the course. The science methods notebook is a formative assessment that lends itself naturally to metacognitive questions. Students write about how their own science ideas are developing and what difficulties in learning they have. We can also ask students to be metacognitive in a summative assessment at the end of the course. For example, in preparation for a teaching portfolio exit interview, an instructor might ask students to pinpoint critical points in their learning during the course, as well as identify areas for continued growth as they enter their professional careers.

It's Your Turn

Think about an assignment that you currently use in your elementary science methods course. *Does it have a metacognitive component?* If not, think about how you could add a part to the assignment that would require your students to think about the progress they are making in thinking like a science teacher.

Scoring and Grading Assignments

There will be many times throughout the course where you will give diagnostic and formative assessments that will not be graded. That is because their purpose is to provide an ongoing account of how students are making sense of course ideas. However, as the course instructor, you will most likely be charged with calculating a grade for each student at the end of the course. One of the purposes of summative assessments is to determine the final grade. Typically you will be grading summative assessments that carry differing amounts of weight in determining the final grade.

The first step in grading an assignment is to develop a method for scoring the assignment. When scoring assessments, three different approaches are possible: norm-referenced, criterion-referenced, and learner-referenced approaches.

- *Norm-referenced.* When assessment data for a given student are compared against a group of other students for the purpose of a score or grade, that is considered norm-referencing. Grading a test on a curve is an example of a norm-referenced assessment.
- *Criterion-referenced.* When assessment data for a given student are compared to a performance standard, then the assessment is criterion-referenced. Scoring rubrics provide a method of criterion-referenced assessment.
- *Learner-referenced.* When we compare a student's performance at a point in time with his/her previous performance, we are using a learner-referenced approach. Portfolio assessment is an example of learner-referenced assessment.

We focus on criterion-referenced scoring, the main form of scoring we use in the elementary methods course. For each major assignment, we create a scoring sheet or a rubric to guide our evaluation of student products.

Scoring Rubrics

A scoring rubric provides a quantitative means of evaluating student work as compared with a defined set of criteria. Typically the instructor creates the rubric at the same time the assignment is designed. However, we also have found it valuable for the methods students themselves to help design a scoring rubric for one or more course assignments.

In designing a scoring rubric, the instructor makes a number of decisions. The first has to do with the overall nature of the rubric—whether it will be holistic or analytic. The holistic rubric defines performance levels by a group of criteria. For example, "An 'A' paper is well organized and easy to follow, makes a clear argument for the importance of inquiry-based instruction, and uses evidence from course readings and course activities to support the argument." Holistic rubrics support overarching judgments about the quality of student work. On the other hand, an analytic rubric breaks down performance into a number of categories. Most of the rubrics we have developed for the elementary science methods course are analytic in nature.

Several steps are involved in creating an analytic scoring rubric.

1. Decide on the criterion categories. These will depend on the learning goals for the assignment. For example, in an assignment in which methods students analyze their interviews about children's science ideas, one goal is that methods students will be able to compare what they learned in the interview to what the research says. Thus, one category of the scoring rubric is "comparison with the research on student misconceptions."
2. Decide on what constitutes excellent performance. Think about what the best response would look like. For example, in the category of "comparison with the research on student misconceptions," excellent performance would be characterized by including comparisons to the research on student misconceptions that are clear and complete, and citing several research papers as evidence.
3. Decide on the number of levels of performance that will be included in the reference. Three or four levels of performance are typical.
4. Decide on what constitutes the other levels of performance, for example, "satisfactory" and "poor" performance. Some instructors like to define the lowest level of performance next, and then fill in with defining other performance level(s).
5. Decide on the number of points to award to the entire product, to this category, and to each performance level.

Table 7.3 displays one row of a scoring rubric developed for the interview assignment, based on the criterion of "comparison to the research." The row of the rubric defines three levels of performance and assigns scores to each level. The full rubric would include several rows with various performance criteria.

Rubrics can also be developed to score student responses to essay exams. For example, Table 7.4 displays a full scoring rubric for a final exam in which students answered three questions related to a lesson that they had developed as part of their field experience. Rubrics like these assist instructors in evaluating products fairly and efficiently, and help students to see where their work could be improved. Many tools for creating scoring rubrics are available on the Internet (for example, http://www.rubistar.com) and can be useful for newcomers to rubric design.

Instructors sometimes wonder at what point the student should see the scoring rubric. Sometimes we hand out the scoring rubric along with the

Table 7.3 Example Row of a Scoring Rubric for Interview Assignment

Criterion	Excellent performance	Satisfactory performance	Poor performance
Compare and contrast student's ideas to the research on students' misconceptions. (worth 10 points out of 50 total)	Comparisons to the research on student misconceptions are clear and complete. Several research papers cited as evidence. (9–10 points)	Some comparisons to the student misconceptions literature are made, but few research papers cited. (7–8 points)	Comparisons to the student misconceptions literature are weak or non-existent. (6 or fewer points)

assignment description. Other times we include the evaluation criteria in the assignment description, but not the full scoring rubric. We have found that both ways serve to make our expectations for student performance clear to them and to us.

Although we find rubrics useful to both instructors and students, we do not create full rubrics for every graded methods course assignment. Sometimes we opt for a scoring sheet that assigns points based on the parts of the assignment, and allows instructors some leeway in judging performance on each assignment component. In the "Tools" section of this chapter, we have included two examples of scoring tools we have used in the elementary science methods course. The first example is a scoring sheet based on assignment components and evaluation criteria. The second example is a scoring rubric with levels of performance developed by the methods students in collaboration with their instructor.

Determining Final Grades

Typically the elementary science methods course instructor decides how to determine final grades before the course begins, and indicates this in the course syllabus (see Chapter 5). The syllabus will include a description of the graded assignments and the percent of the total grade that each assignment is worth (see Table 7.5). The decision about how to weight the various assignments relative to each other will take into consideration the difficulty and importance of the assignment.

The syllabus in Table 7.5 also displays the grading scale for the methods course. The grading scale clarifies that the course is criterion-referenced, and that performance on the various assignments will be calculated to determine the final grade. In a norm-referenced course, student performance relative to each other would be part of determining grades, and some curving system would be used.

Table 7.4 Example of a Full Scoring Rubric for Final Exam

Examination essay question	Excellent 81–100%	Satisfactory 60–80%	Poor <60%
A. Describe how you think your lesson is consistent with one or more ideas about the Nature of Science. (5 pts.)	1. Clear and accurate description of one or more components of NOS is provided. *and* 2. An accurate connection is made between the lesson and NOS. (4–5)	1. Unclear and inaccurate description of one or more components of NOS is provided. *or* 2. An inadequate connection is made between the lesson and NOS. (3–4)	1. Unclear or inaccurate description of one or more components of NOS is provided. *and* 2. An inaccurate connection is made between the lesson and NOS. (0–3)
B. Describe how you think your lesson incorporates one or more of the five essential features of inquiry. (5 pts.)	1. One or more of the five essential features of inquiry are clearly described. *and* 2. An accurate connection is made between inquiry and the lesson. (4–5)	1. One or more of the five essential features of inquiry are unclearly described. *or* 2. The connection made between inquiry and the lesson is not accurate. (3–4)	1. One or more of the five essential features of inquiry are unclearly described *and* 2. The connection between inquiry and the lesson is not accurate. (0–3)
C. Describe how your lessons follow a 5E instructional sequence. Which E(s) does your lesson fit? Describe this fit. (5 pts.)	1. The 5E instructional sequence is accurately described. *and* 2. A clear and accurate connection and rationale is made between the two lessons and the 5E. (4–5)	1. The 5E instructional sequence is described incorrectly. *or* 2. A clear and accurate connection or a rationale is not made between the two lessons and the 5E. (3–4)	1. The 5E instructional sequence is described incorrectly. *and* 2. A clear and accurate connection and rationale is not made between the two lessons and the 5E. (0–3)

Final Score

Table 7.5 Excerpt from Elementary Science Methods Course Syllabus Related to Grading

Assignments
1. *Attendance, participation, and reflections.* Because of the interactive nature of this course, regular attendance and high quality participation is expected. You will be asked to reflect in writing about course activities and readings throughout the course. *(15% of total grade)*
2. *Moon watch notebook.* To help you think about teaching science, it is important for you to think about yourself as a science learner. In this activity you will make observations of the moon, invent explanations that fit your observations, and reflect on the experience. *(15% of total grade)*
3. *Small group science lesson.* This activity will allow you to "get your feet wet" in terms of science teaching by planning a short lesson and carrying it out with a small group of students at your field experience site. *(10% of total grade)*
4. *Interview.* It is critical in your science teaching that you understand your students' ideas as you design learning experiences. One way to assess student ideas is through individual interviews. In this assignment you will design an interview, interview two children, record and transcribe the interview, analyze the findings, and compare your findings with other students. *(15% of total grade)*
5. *Team unit plan.* When you begin classroom science teaching, a major task you will face is developing curriculum in the form of science units. For this major course assignment, you will work with a team to design and enact a series of science lessons at a local elementary school. *(20% of total grade)*
6. *Final Paper.* In lieu of a final examination, the final paper will help you to synthesize your ideas about elementary science teaching and learning. You will become a member of a Science Curriculum Adoption Committee and write an instructional materials analysis and evaluation. *(25% of total grade)*

Grading
Because of the interactive nature of this course, regular attendance is expected (see above). Absences, excused or unexcused, could result in the lowering of your final grade. All assignments will be graded using criterion-referenced methods. That is, they will be scored against a specific set of standards. Your final grade will be calculated based on a percentage of total possible points: 90–100% = A; 80–89% = B; 70–79% = C; 60–69% = D.

It's Your Turn

Develop a scoring rubric for one of your existing course assignments. Think about what would constitute excellent performance on this assignment. *What are the components of this excellent performance?* Use these ideas to expand your rubric to include criteria for average and unsatisfactory performance. Revisit your assignment to be sure that your rubric addresses expectations in the assignment and that your assignment describes the evaluation criteria in your rubric.

Putting it All Together: Designing a Methods Course Assignment

Designing major course assignments that document student learning of important course concepts and provide data for the final course grade is a major component of planning the elementary science methods course. Providing feedback and scoring these assignments is a major component of course implementation. When we clearly communicate our expectations for course assignments, we find that student products are better and easier to score. Thus we provide our students detailed course descriptions for each major course assignment. In this section we include a brief guide to designing assignments accompanied by an actual example from our teaching.

Step 1

We believe that students perform better when they understand the reason for the assignment. Thus we begin each assignment description with a purpose or rationale, linked to course goals, and a brief overview of the assignment. One assignment we have used is the Inquiry Station Assignment, in which methods students use what they have learned about productive questions (Harlen, 2001) to design a science learning station for K-2 students in a field experience. We begin the assignment description by setting out the following purpose:

Inquiry Station Assignment

Purpose

One way to promote more science doing and thinking in your classroom is through the use of inquiry stations. In this assignment you and your team will put together what you know about productive questions and developmentally appropriate science (from the science standards) to create a thematic set of science inquiry stations, chockfull of opportunities for science exploration. You will lead your station activities for small groups of students at your field experience site.

Step 2

We have found that when we provide details about our expectations, students are more confident and perform better. Therefore we like to include detailed procedures for each assignment. For example, for the Inquiry Station Assignment, the following set of procedures guides methods students' work:

> **Inquiry Station Assignment**
>
> *Procedure*
>
> 1. Work with a team of methods course students to develop a set of science inquiry stations for your field experience classroom. Design your stations around a particular science topic (e.g., force and motion, erosion, life cycles). Ask your classroom teacher which science topic she would prefer your team to work on. As a team think of an overarching question that will guide your stations (see *examples* below). (*Note*: have your guiding question approved by your methods course instructor before proceeding).
>
>   *Examples of guiding questions:*
>
>   *Physical Science*. How do mirrors reflect light? How can I make different sounds?
>   *Earth Science*. How does the sun move throughout the day? What are different soils like?
>   *Life Science*. How do earthworms behave? How do plants and animals interact?
>
> 2. Check to see that your guiding question relates to science standards at your grade level. Make sure the members of your team understand the science concepts involved in your topic. Find resources (on the Internet and in the Curriculum Library) with ideas for activities about your topic.
> 3. Each team member should create one inquiry station that addresses your team's guiding question. You will need to design open-ended activities, create a set of productive questions (Harlen, 2001), and compile all equipment needed at your station to engage children in doing and thinking using the station materials as a starting point. Write at least two productive questions for each type, including reasoning. Create a poster board display for your station.
> 4. At the school site the class will be divided into as many groups as there are members of your team. These groups will rotate through each of the stations, after a brief introduction by the team. You will be in charge of facilitating the students at your station, using the productive questions you developed.

Step 3

Along with these detailed procedures, we make sure that students understand the format we expect for the final product. For example, in the Inquiry Station Assignment, we give students the following guidance about the format of the product they will turn in:

> **Inquiry Station Assignment**
>
> *Final product*
>
> Your final product will be a written plan for your individual inquiry station. The plan should be specific and detailed such that another teacher could use it. This plan should include the following components:
>
> - Title, in the form of a guiding question (same for all team members).
> - Grade level (same for all team members).
> - Purpose.
> - Materials list.
> - Photo of your station.
> - Conceptual background:
>   - What basic science concepts underlie the station activities?
>   - What science standards do your station activities address?
> - Overall lesson description for your station.
> - List of productive questions, labeled by type.
> - Brief description of how students interacted with the station—which questions and activities were most effective in engaging students in thinking and doing?
> - References: fiction book(s), nonfiction book(s), and URL(s) with description.

Step 4

We believe that students should understand the criteria we will use to evaluate the assignment. Therefore, with each assignment description, we include a set of questions that students can use to evaluate their final product before turning it in. These questions form the basis of the scoring rubric (see previous section) that we will use to grade their work.

> **Inquiry Station Assignment**
>
> *Evaluation*
>
> Your final product will be evaluated using the following criteria:
>
> - Completeness (Did you include each component as listed in the assignment sheet?).
> - Thoroughness (Did you address each component thoughtfully in light of what we have discussed about student engagement?).
> - Appropriateness (Was your station appropriate for the topic and the grade level?).
> - Use of Productive Questions (Did you include examples of each type of productive question and were they correctly classified?).
> - Clearly written (Was your final paper clearly organized, with minimal spelling, grammar, and punctuation errors?).

Step 5

We develop the scoring guide for the assignment at the same time we develop the assignment description, aligning the evaluation criteria in the assignment description with the actual scoring criteria. In this case, we opted for a scoring sheet rather than a scoring rubric.

Inquiry Station Assignment

Scoring Sheet

Name _____

Score (20 pts. possible) _____ *{This assignment is worth 10% of your total grade.}*

Completeness/Thoroughness. Are each of the following components present and thoughtfully completed? (4 pts.)
 _____ Title (in the form of a guiding question) and grade level
 _____ Purpose
 _____ Materials list
 _____ References: fiction, nonfiction, and URL

Quality of the Station Activity. Does your activity have a conceptual background that is age appropriate and does it demonstrate what we know about best ways to engage students with phenomena? (6 pts.)
 _____ Description of science concept(s) involved and connections to age appropriate science standards
 _____ Inquiry-oriented station activity (not a recipe)
 _____ Integration of science writing

Productive Questions. Are there at least two questions of each type and are they correctly classified? (10 pts.)
 _____ Attention-focusing
 _____ Measuring and counting
 _____ Comparison
 _____ Action
 _____ Problem-posing
 _____ Reasoning (how and why)

Over time, we improve our skills at designing clear assignment descriptions that guide students through the assessment task. We have found that our first attempts at describing assignments are sometimes confusing to students. Sometimes, when we obtain their final products, we find that our directions were not completely aligned with our expectations. Each time we teach the elementary methods course, we improve our assessments and clarify them for students through the assignment descriptions. We encourage you to experiment with your assessments, reflect on your own progress as an instructor, and continuously find ways to improve your work.

> **It's Your Turn**
>
> Design a new assignment for your class linked to a subset of your course goals. Think about how the assignment connects to the jobs of science teaching, and make sure those jobs are evident in the assignment. Use the assignment design section of this chapter to include all of the components of an effective assignment.

Conclusion

When we teach future teachers, our teaching is on display. Our students learn from what we do, not merely from what we say or what they read. Thus we must take care to model best practices in the methods course. Regarding assessment, our continuing challenge as elementary science methods course instructors is to employ the purposes and principles of effective assessment to design assessment tasks that inform us about what students know and can do. We use this evidence to inform our instruction, and to provide feedback to students that is aligned with our course goals.

> Rhonda and I met several times to discuss assessment in our methods courses since our initial encounter at the copy machine. I shared a number of assignments and strategies with her, as well as examples of student products. Rhonda recognized that my assessment processes help me understand what my students know and don't know. However, rather than throwing out everything with which she is comfortable, she decided to revise her course assessment methods over time. She retained her mid-term exam, but replaced her final exam with an application exercise in which she asked students to evaluate a unit from an elementary science curriculum using principles for science teaching and learning they had derived in the course throughout the semester. We worked together to construct a scoring rubric for the assignment. Now, several semesters later, Rhonda has revised the assignment and the rubric to align better with her course goals and instruction. According to Rhonda, the students have placed their work on this assignment in their certification portfolios for use in job interview settings.

Tools for Teaching Elementary Science Methods

Summative Assessment Assignment and Scoring Tool: Science Instructional Materials Analysis

Purpose

An important learning task in any course is to synthesize the learning experience—to tie together all of the pieces and construct a coherent meaning for yourself. You should consider this assignment as the final exam for this course. The paper you write will prove to be useful as you put together your teaching portfolio, complete job applications, and answer job interview questions.

Procedure

You have been selected to be on the Science Curriculum Adoption Committee in the Science for Striving for Excellence School District, representing grade 3. One of the curriculum programs your committee is examining is XXX. You have been given the job of critiquing the third grade "States of Matter" unit from this curriculum and giving a report to the chair of the committee.

First you will use parts of the curriculum analysis tool—which we have used this semester in class—to analyze the components of the science curriculum module. You will then write a paper based on this analysis and make a recommendation to the adoption committee.

Data Collection: Analysis Clusters

The following analysis clusters represent the major topics from this course. Using the criteria on the analysis tool provided, find instances of the clusters in the curriculum module. The more specific you are with citing your instances, the easier it will be to write your paper. More information on curriculum analysis can be found at the AAAS website (http://www.project2061.org).

A. Engaging Students with Phenomena
B. Providing a Sense of Purpose
C. Promoting Student Thinking about Phenomena
D. Taking Account of Student Ideas
E. Assessing Progress
F. Enhancing the Social Dimensions of the Science Classroom.

Data Interpretation: Writing your Paper
1. Your paper should begin with an *introduction*. In it you can introduce and briefly describe the organization of this curriculum module.

2. Next you should write a *section for each of the analysis clusters* in which you:
   a. *Describe* the cluster. What do the criteria mean in terms of good science teaching? Link your description to course readings and activities.
   b. *Cite specific examples* from the curriculum that address or do not address each criterion within the cluster. Explain why you chose those examples to represent each criterion.
   c. *Evaluate* the curriculum for that cluster. Is the evidence sufficient in support of the cluster? Does the curriculum do a good job of addressing the cluster? How does it compare to course activities and readings?
3. Your paper should conclude with a *summary recommendation*. Using the data you have collected, provide a recommendation as to whether or not the Striving for Excellence School District should adopt the curriculum based on these criteria. This recommendation does not have to be all or none. For example, you could decide to adopt with recommendations for adapting the curriculum in the areas of weakness that you noted.

Specifications

1. Your paper will be easiest to read if you use subheadings to mark each time you begin discussing a new cluster.
2. Your critique should contain your evaluation, backed up by evidence. The evidence should be *specific* examples from the curriculum. Use course books, the entire reading packet, field experiences, course videos and activities, and your written reflections throughout the course to support your critique.
3. When you give evidence from course readings, you must give credit to the reference. See "How to Cite Course Readings in Your Writing" in the syllabus.
4. Your critique should be typed, double spaced, and 10–12 pages (not counting title page or bibliography). Please number the pages.
5. Papers are due in class on the last period of the semester.

Evaluation

This assignment will comprise 25% of your final grade. It will be evaluated based upon the following criteria:

- Did you follow the procedures as described?
- Did you include a critique of each cluster?
- Did your critique include Descriptions, Examples, and Evaluation for each cluster?
- Did you cite examples from the curriculum for each cluster critique?

- Did you support your critique with a variety of course readings and experiences?
- Did you include an introduction?
- Did you include a summary recommendation with suggestions for adapting the curriculum?
- Is your final product logically organized, understandable, and well written (spelling, grammar, etc.)?

Science Instructional Materials Analysis Scoring Sheet

Name _____ Score (50 pts. possible) _____

_____ Introductory paragraph (2 pts.)

- Cluster A: Engaging Students with Phenomena (6 pts.)
  _____ description (component fully described and linked to course readings and activities)
  _____ examples (cited specific examples from the curriculum to support/refute criteria)
  _____ evaluation (your critique provided and compared to course readings and activities)
- Cluster B: Providing a Sense of Purpose (6 pts.)
  _____ description (component fully described and linked to course readings and activities)
  _____ examples (cited specific examples from the curriculum to support/refute criteria)
  _____ evaluation (your critique provided and compared to course readings and activities)
- Cluster C: Promoting Student Thinking about Phenomena (6 pts.)
  _____ description (component fully described and linked to course readings and activities)
  _____ examples (cited specific examples from the curriculum to support/refute criteria)
  _____ evaluation (your critique provided and compared to course readings and activities)
- Cluster D: Taking Account of Student Ideas (6 pts.)
  _____ description (component fully described and linked to course readings and activities)
  _____ examples (cited specific examples from the curriculum to support/refute criteria)
  _____ evaluation (your critique provided and compared to course readings and activities)
- Cluster E: Assessing Progress (6 pts.)
  _____ description (component fully described and linked to course readings and activities)

_____ examples (cited specific examples from the curriculum to support/refute criteria)
_____ evaluation (your critique provided and compared to course readings and activities)
- Cluster F: Enhancing the Social Dimensions of the Science Classroom (6 pts.)
  _____ description (component fully described and linked to course readings and activities)
  _____ examples (cited specific examples from the curriculum to support/refute criteria)
  _____ evaluation (your critique provided and compared to course readings and activities)
  _____ Summary recommendation (6 pts.)

Other considerations (6 pts.)
_____ well organized and understandable: logical argument, minimal use of jargon, specific examples
_____ appropriate use of citations and references
_____ well written: grammar, spelling, mechanics

Summative Assessment Assignment and Scoring Tool: The Science Teaching Portfolio

Rationale

Each of you came to the course with your own unique experiences and ideas about science and science teaching and learning. Over the course of the semester, you have been presented with new ideas and perspectives, which have further shaped your ideas. What you have "learned" this semester, then, consists of the ways in which you have grown in your thinking and/or changed your ideas. Summative assessment in the course is intended to allow you multiple and varied ways to communicate your individual growth and progress this semester through a portfolio.

The science teaching portfolio is an edited collection of evidence and reflections representing your progress toward achieving the course goals. It provides the basis for self-assessment of your learning, and my evaluation of your progress this semester. Note: the portfolio is designed to capture your progress over time, and is not just a snapshot of where you are at the end of the course.

Procedure and Specifications

The following guidelines are designed so that there are multiple acceptable ways through which to satisfy the criteria. Please pay close attention to these guidelines, as well as the evaluation rubric as you prepare to your portfolio. The final portfolio should include:

1. An organization schema to assist readers in navigating your portfolio.
2. Artifacts/evidence that you have made progress toward each of the six course goals.
3. A statement of your beliefs about science teaching and learning.

Organizational Schema
Your organizational schema should not only be designed such that your portfolio is easily navigable (e.g., by use of tabbed dividers, color coding, page numbers, etc.) but so that the contents are arranged in meaningful ways. You may group your evidence chronologically, according to the six course goals, by theme, etc. A table of contents is required.

Evidence
The choice of what and how much evidence to include in your portfolio is yours. Choose artifacts that are most meaningful to you. Throughout the semester, you have completed course assignments and participated in numerous activities that have not been graded or collected. Many students rely on these as artifacts, in addition to graded assignments. Artifacts that you have *created* (your own work) often provide stronger evidence about your knowledge and abilities than those that you have merely *collected* (resources you've gathered). Each piece of evidence should be identified (What is the artifact? When was it created? In what context?). Additionally, a rationale for its inclusion should be provided. This rationale should indicate what you feel the evidence demonstrates about your progress.

Science Teaching Statement
The statement of your beliefs about science teaching should be clear and succinct. (No more than 10 pages double-spaced; 12-point Times New Roman; 1-inch margins.) It should reflect not only your personal beliefs, but also your professional opinion, which is drawn from sources other than your own experiences. Provide supporting citations from the course readings and reference appropriate evidence from your portfolio that illustrate your beliefs. Your statement should synthesize the progress you've made this semester, and what the work in your portfolio demonstrates about you as a teacher of elementary science.

At a minimum, your statement should address these questions:

- What is science?
- Why is it important for elementary students to learn science?
- How do elementary students learn science?
- How should science be taught?

Evaluation
Your portfolio evaluation will be based on the scoring guide attached to this assignment.

Science Teaching Portfolio Scoring Rubric
Name: _____   Conference Date/Time: _____

Analytic evaluation	Scoring guide:	
	| 1 = Exemplary, 2 = Proficient, 3 = Adequate, 4 = Unsatisfactory	
Organization	1	Easily navigable; reader-friendly. Table of contents thorough and complete. Professional in appearance with logical sequence evident.
	2	Well organized and professional in appearance. The table of contents is complete.
	3	Some attempt at organization, but it is difficult for the reader to locate individual portfolio contents easily.
	4	Lacks organizational schema; navigation of portfolio is difficult.
Quality of evidence	1	Portfolio contains multiple pieces of evidence to support progress toward *each* course goal. The link between course goals and evidence is strong. Diverse, represents both depth and breadth of knowledge. Evidence documents professional growth and change.
	2	Portfolio contains evidence to support progress toward *each* course goal. Evidence relates to each of the course goals. Variety of evidence presented. Professional growth evident throughout the portfolio.
	3	Portfolio contains evidence to support progress toward *each* course goal. Evidence is only loosely connected to course goals, or the connection is not made explicit. Professional growth is evident in most areas.
	4	Portfolio does not contain evidence to support progress toward each course goal. Evidence is not connected to course goals. Growth and progress not evident.
Teaching statement	1	Articulate, well developed, logically organized, and clearly expressed. Student indicates a firm commitment to his/her educational philosophy with supportive reasoning. Based on professional, rather than personal knowledge. Reflects an understanding of science teaching and learning, consistent with the overall portfolio and evidence.
	2	Logically organized; beliefs are expressed and supported with examples. Writing reflects an understanding of science teaching and learning, consistent with the overall portfolio and evidence.
	3	Statement is coherent; support is general rather than specific. The student fails to connect beliefs to the portfolio evidence or his/her beliefs about science teaching and learning are not consistent with the overall portfolio and evidence.
	4	Writing is disorganized and/or reasoning is not presented in a logical sequence. Insufficient depth and elaboration of ideas about science teaching and learning. Written statement does not coincide with portfolio evidence presented.

Instructor's comments:
Portfolio final grade: _____

Note
Adapted from Moseley (2000).

More to Explore

For more ideas about assessment in science teaching:

Abell, S. K., & Volkmann, M. J. (2006). *Seamless assessment in science: A guidebook for elementary and middle school teachers*. Portsmouth, NH: Heinemann.
Bell, B. (2007). Classroom assessment of science learning. In S. K. Abell & N. G. Lederman (Eds.), *Handbook of research on science education* (pp. 965–1006). Mahwah, NJ: Laurence Erlbaum.
Black, P. (1993). Formative and summative assessment by teachers. *Studies in Science Education, 21*, 49–97.
National Research Council. (2001). *Classroom assessment and the national science education standards*. Washington, DC: National Academy Press.
Smith, S. R., & Abell, S. K. (2008). Assessing and addressing student science ideas. *Science and Children, 45*(7), 722–773.

For more ideas about assessment in the college classroom:

Angelo, T. A., & Cross, K. P. (1993). *Classroom assessment techniques: A handbook for college teachers* (2nd ed.). San Francisco: Jossey-Bass.
Bean, J. C. (1996). *Engaging ideas: The professor's guide to integrating writing, critical thinking, and active learning in the classroom*. San Francisco: Jossey-Bass.
Hanuscin, D., Richard, M., Chandrasekhar, M., Corman, A., & Lapilli, C. (2007). Collaborative action research to improve classroom assessment in an introductory physics course for teachers. *Journal of Physics Teacher Education Online, 4*(2), 16–20.

References

Abell, S. K. (1992). Helping science methods students construct meaning from text. *Journal of Science Teacher Education, 3*, 11–15.
Abell, S. K. (2007). On reading in science. *Science and Children, 45*(3), 56–57.
Abell, S. K. (2009). Thinking about thinking in science class. *Science and Children, 46*(6), 56–57.
Abell, S. K., Anderson, G., & Chezem, J. (2000). Science as argument and explanation: Inquiring into concepts of sound in third grade. In J. Minstrell & E. van Zee (Eds.), *Inquiring into inquiry learning and teaching in science* (pp. 65–79). Washington, DC: American Association for the Advancement of Science.
Abell, S. K., George, M. D., & Martini, M. (2002). The moon investigation: Instructional strategies for elementary science methods. *Journal of Science Teacher Education, 13*, 85–100.
Abell, S. K., & Volkmann, M. J. (2006). *Seamless assessment in science: A guidebook for elementary and middle school teachers*. Portsmouth, NH: Heinemann.
Allen, P. (1982). *Who sank the boat?* New York: Coward-McCann, Inc.
Anderson, L. W., Krathwohl, D. R., Airasian, P. W., Cruikshank, K. A., Mayer, R. E., Pinrich, P. R., et al. (2001). In L. W. Anderson & D. R. Krathwohl (Eds.), *A taxonomy for learning, teaching, and assessing: A revision of Bloom's taxonomy of educational objectives*. New York: Longman.

Angelo, T. A., & Cross, K. P. (1993). *Classroom assessment techniques: A handbook for college teachers* (2nd ed.). San Francisco: Jossey-Bass.

Banta, T. W., Lund, J. P., Black, K. E., & Oblander, F. W. (1996). *Assessment in practice: Putting principles to work on college campuses.* San Francisco: Jossey-Bass.

Bean, J. C. (1996). *Engaging ideas: The professor's guide to integrating writing, critical thinking, and active learning in the classroom.* San Francisco: Jossey-Bass.

Bransford, J. D., Brown, A. L., & Cocking, R. R. (Eds.). (1999). *How people learn: Brain, mind, experience, and school.* Washington, DC: National Academy Press.

Chambers, D. W. (1983). Stereotypical images of the scientist: The Draw-A-Scientist Test. *Science Education, 67,* 255–265.

Cobern, W. W., & Loving, C. C. (1998). The card exchange: Introducing the philosophy of science. In W. McComas (Ed.), *The nature of science in science education: Rationales and strategies* (pp. 73–82). Dordrecht, the Netherlands: Kluwer.

Finson, K. D., Beaver, J. B., & Cramond, B. L. (1995). Development of and field-test of a checklist for the draw-a-scientist test. *School Science and Mathematics, 95,* 195–205.

Friedrichsen, P., & Dana, T. (2003). Using a card sorting task to elicit and clarify science teaching orientations. *Journal of Science Teacher Education, 14,* 291–301.

Hanuscin, D., & Lee, M. H. (2008). Using the learning cycle as a model for teaching the learning cycle to preservice elementary teachers. *Journal of Elementary Science Education, 20*(2), 51–66.

Harlen, W. (2001). *Primary science: Taking the plunge* (2nd ed.). Portsmouth, NH: Heinemann.

Hubbard, P., & Abell, S. (2005). Setting sail or missing the boat: Comparing the beliefs of preservice elementary teachers with and without an inquiry-based physics course. *Journal of Science Teacher Education, 16,* 5–25.

Keeley, P. (2008). *Science formative assessments: 75 practical strategies for linking assessment, instruction, and learning.* Thousand Oaks, CA: Corwin Press and NSTA.

Koch, J. (1990). The science autobiography. *Science and Children, 28*(2), 42–43.

Lederman, N. G., Abd-El-Khalick, F., Bell, R. L., & Schwartz, R. S. (2002). Views of nature of science questionnaire: Toward valid and meaningful assessment of learners' conceptions of nature of science. *Journal of Research in Science Teaching, 39,* 497–521.

Lee, M. H., & Hanuscin, D. L. (2008). A (mis)understanding of astronomical proportions? *Science and Children, 46*(1), 60–61.

Lortie, D. C. (1975). *Schoolteacher: A sociological study.* Chicago: University of Chicago Press.

Loughran, J., Berry, A., & Mulhall, P. (2006). *Understanding and developing science teachers' pedagogical content knowledge.* Rotterdam, the Netherlands: Sense Publishers.

Moseley, C. (2000). Standards direct pre-service teacher portfolios. *Science and Children, 36*(5), 39–43.

National Research Council. (1999). *Selecting instructional materials.* Washington, DC: National Academy Press.

Novak, J. D. (1998). *Learning, creating and using knowledge: Concept maps as facilitative tools in schools and corporations.* Mahwah, NJ: Erlbaum.

Novak, J. D., & Gowin, D. B. (1984). *Learning how to learn.* London: Cambridge University Press.

Osborne, R., & Freyberg, P. (Eds.). (1985). *Learning in science: The implications of children's science.* Portsmouth, NH: Heinemann.

Thomas, J. A., Pederson, J. E., & Finson, K. (2001). Validating the Draw-A-Science-Teacher-Test Checklist (DASTT-C): Exploring mental models and teacher beliefs. *Journal of Science Teacher Education, 12,* 295–310.

van der Valk, A. E., & Broekman, H. (1999). The lesson preparation method: A way of investigating pre-service teachers' pedagogical content knowledge. *European Journal of Teacher Education, 22,* 11–22.

Wiggins, G. P., & McTighe, J. (2005). *Understanding by design* (2nd ed.). Alexandria, VA: Association for Supervision and Curriculum Development.

Chapter 8

Field Experiences in Elementary Science Methods[1]

> "My science methods students say great things in class and on written assignments. I really think they are going to be terrific science teachers!" Carl exclaimed as we passed in the hallway. Carl's claim made me wonder if responses on methods course assignments were enough to signify future teaching excellence. "Don't students also need the knowledge and skill that comes with practice?" I wondered. The next day when I saw Carl, I proposed that we add an opportunity to our methods course for students to work with real students in real classroom settings. "This might help them put their developing ideas about science teaching and learning to the test. "Interesting, but it sounds like a logistical nightmare," Carl responded. "However, I'm game if you are."

A course on teaching science in the elementary school that lacks opportunity to practice with real elementary students and a more experienced teacher is no more complete than a driver education course where the student never gets behind the wheel. In becoming a teacher of science, students need to understand the principles of practice (Shulman, 1986) that can be learned in on-campus settings, but they also need to solidify their understanding with opportunities to apply these principles in real situations. In the classroom, preservice teachers reflect-in-action (Schön, 1983, 1987, 1991), making decisions on the spot about everything from what question to ask next to how to distribute materials to how to respond to a student who is not participating. No methods course in the world can prepare future teachers for the level of synthesis that takes

---

1. Parts of this chapter were adapted from: Abell, S. K. (2006).

place in classroom settings. For these reasons, we believe that the field experience must be a component of elementary science teacher preparation. The purposes of this chapter are: (1) to discuss the benefits and challenges of field experiences that are associated with the elementary science methods course, (2) to describe several models of field experiences we have experienced, and (3) to illustrate the kinds of reflection assignments that can be used to help students learn from field experiences. We believe that the potential benefits of field experiences are tremendous and the challenges surmountable.

Benefits of Field Experiences

Current learning theories propose that learning is situated in authentic contexts that allow learners to participate in communities of practice (Brown, Collins, & Duguid, 1989; Lave, 1988; Lave & Wenger, 1991; Rogoff, 1990). Druckman and Bjork (1994) characterized four common aspects of the situated approach:

1. The actions we take are grounded in the concrete situations. One's knowledge is specific to the situation at hand.
2. Tasks are not accomplished solely by the application of rules or principles. Performance in a particular context by necessity goes beyond rule-based thinking.
3. Learning happens through doing. The most effective training occurs in situations that closely match prior experience and involve apprenticeships.
4. Performance is social. Thus we need to understand how social interactions affect performance.

One implication of the situated approach is that learners learn best in authentic apprenticeship contexts alongside veterans (Lave & Wenger, 1991; Rogoff, 1990).

The situated learning perspective can be applied to the context of teacher learning (see Putnam & Borko, 2000). The performance environment is the elementary classroom, where students of teaching take part in apprenticeships in which they join, peripherally at first, and more fully as time goes on, the community of practice of teaching. Their actions are grounded in the concrete classroom situations that occur naturally. Knowles, Cole, and Presswood (1994) proposed a spiral apprenticeship model of teacher learning through field experience. In their model, "experience is the foundation for learning" (p. 10), supported by information gathering, reflection, analysis, and informed action. One criticism of teacher education is that it often creates a theory/practice gap between the "ivory tower" of the university and the "real world" of the classroom. The apprenticeship model attempts to link theory and practice in a situated context. Thus, the situated learning perspective provides theoretical support for including field experiences in the elementary science methods course.

Although the theoretical support for the field experience is strong, the research-based support is limited (Abell, 2006; Clift & Brady, 2005; McIntyre, Byrd, & Foxx, 1996; Wilson, Floden, & Ferrini-Mundy, 2002). The research on field experiences supports a few important conclusions:

- prospective teachers sometimes receive conflicting messages from methods instructors, partner teachers, and field supervisors;
- even when field placements reinforce ideas from the methods course, prospective teachers may resist changing their beliefs or practices;
- prospective teachers find it difficult to translate ideas from the methods course into the classroom (Clift & Brady).

These conclusions confirm that field experiences must be carefully planned, implemented, and evaluated to be effective. Field experiences are important to our methods students and to our school-based colleagues; we must work hard to design field experiences that take these conclusions into consideration.

What can field experiences provide that traditional university-based coursework cannot? Experience teaches us that the field provides opportunities for prospective teachers to:

- observe and listen to elementary students, understand how they reason scientifically, and even reshape expectations regarding elementary students' science abilities;
- operationalize inquiry-based approaches learned in the science methods course;
- practice specific science teaching routines, strategies, and approaches (e.g., materials distribution, questioning, cooperative learning, demonstrations, learning cycle) and see how students respond;
- revise beliefs about what works in elementary science classrooms;
- understand the diverse roles of the science teacher;
- reflect in and on action (Schön, 1983, 1987, 1991).

Field experiences also can have indirect outcomes for other stakeholders by providing:

- increased opportunities for elementary students to learn science with understanding and increase their interest in science;

- professional development for veteran teachers when preservice teachers bring new teaching approaches and strategies into their classrooms;
- strengthened ties between schools and universities;
- positive public relations for colleges of education that want to demonstrate their contributions to the community.

However, many factors conspire to lessen the effectiveness of field experiences or decrease the probability that they will occur at all. We discuss these factors in the next section.

> **It's Your Turn**
>
> *What are your reasons for wanting to include a field experience component with your elementary science methods course?*
> *What challenges do you foresee?*

Addressing Challenges to the Science Methods Field Experience

During our years of teaching future elementary teachers of science, we have witnessed many challenges to the science methods field experience. As Carl said in the introductory vignette, setting up a field experience for the science methods course can be a logistical nightmare. However, we believe that the benefits far outweigh these constraints. The proactive science methods instructor can anticipate the constraints and address them in a variety of ways. In this section we share some of the challenges to the field experience, and some possible solutions. Table 8.1 summarizes these challenges and solutions. Later in the chapter we illustrate these solutions with some examples from our teaching.

Quality Field Placements

The situated learning perspective creates a major challenge for field experiences in science teacher preparation. In our methods courses, we help preservice teachers think reflectively about best practice in science teaching, and we familiarize them with the latest policy documents. We want them to observe, create, and enact reform-minded practice in real elementary classrooms, because we believe the field experience is one form of legitimate peripheral participation (Lave & Wenger, 1991). However, the kinds of classrooms in which we would like our students to serve their apprenticeships are in limited supply. One solution to this constraint is to use video and written cases in which exemplary teachers teach science in reform-minded ways. Another solution involves finding field placements outside the regular science classroom in informal settings such as zoos, science museums, and science clubs where preservice teachers can interact with students as they engage in science. Yet another solution is to build partnerships with teachers and schools over time in which classroom teachers who participate in university-sponsored science professional development form a cadre of potential partners for field experiences.

Scheduling

Once we find enough quality sites, then we face the challenge of scheduling. Most elementary teachers teach science for only a small fraction of the day (if at all), and/or only on certain days at certain times. University students, on the other hand, take courses throughout the day and week, their course and work schedules tend to be unyielding, and their science methods instructors require that field-based assignments be completed on a given timeline. We have coped with the scheduling problem in a number of ways. Using videocases that are always available, or informal science education settings that have more flexible schedules are ways to solve the challenge of scheduling. Another solution is limiting the number of field visits in a given semester and inviting partner teachers who are open those days and times. The most elaborate solution is to build the field experience into the course schedule in addition to the on-campus meetings times, at times when local teachers are teaching science. Students then register for the field component as well as for the class.

Communication

Many of the problems that occur in designing and implementing field experiences arise from incomplete communication among the stakeholders—course instructors, university supervisors, partner teachers, and preservice teachers. The course instructor is responsible for ensuring that all parties know about expectations of the field experience, schedules, and student performance. Regular email and/or website communications before, during and after the field experience are an important part of the solution. Including written expectations for both the preservice teacher and the partner teacher in these communications helps keep everyone on track. If a number of teachers from the same building participate semester after semester, asking one of them to serve as the building liaison can streamline the communications process.

Another communication issue can occur among the instructors in the elementary education program. Oftentimes field experiences are part of the literacy, mathematics, and social studies components of the program.

Working together to find effective and efficient ways to provide field experiences across the various content methods courses can be a major programmatic challenge. We have found that sometimes literacy and mathematics are privileged in the field set up, mirroring the status quo in elementary schools where science and social studies often receive less attention. One way to meet this challenge is regular meetings among the elementary education faculty in which issues can be aired and solutions brainstormed.

Assessment

Students have many questions about field experiences, often tied to assessment. Major questions include how they will be supervised and by whom, what kinds of feedback they will receive, and how their grades will be determined and by whom. What we expect of prospective teachers in a field experience and how we communicate this clearly to them, to other instructors, and to supervisors is key to the success of field experiences. When field supervisors provide feedback to methods course students, they must keep in mind that they are dealing with novices—it is unreasonable to expect them to perform like seasoned veterans, but we can expect them to learn and improve. Because teaching science involves risk-taking, we should make our feedback non-threatening as well as informative. In the "Tools" section of this chapter, we include an example of a feedback form that we have used in the field experience. Partner teachers and university supervisors find that the form provides a guiding structure for their comments, as well as freedom to make comments that fit the context. Methods students find such formative feedback (see Chapter 7) helpful.

Learning from Experience

When our methods students are out in the field teaching science in small and large group settings alongside a veteran teacher, we want to know what they are learning and how they are learning it. This knowledge will help us facilitate their learning. We know from experience that practice can lead to improvement. For example, the more we rode our bicycles as children, the better we got. We figured out on our own how to make sharp turns, navigate road hazards, and make it up hills. However, practice does not automatically lead to expertise. For example, some of us practiced tennis for years, but never became worthy of an above average opponent. In terms of the science teaching field experience, the message is that practice alone does not lead necessarily to learning or improved teaching. Munby and Russell (1994) explained the issue of teacher learning in field settings as one of authority, and raised the question: how can teachers learn from the authority of experience? One useful strategy is to place methods students in teams in their field experiences. This creates a social setting where interns can plan and carry out instruction together, thus providing opportunities for increased learning. We have found that the distributed expertise present in these groups helps preservice teachers be more successful than they would be if working alone. For example, some students understand the science better, others have more experience working with children, and others are better equipped to use inquiry-based strategies. Second, these future teachers will learn more from their experience if they are involved in reflection activities. Working in teams with other methods students provides a ready-made audience for debriefing, and hearing multiple perspectives leads to a more robust understanding of problems of practice. In addition, science educators can create the expectation for individual reflection through written assignments. Later in this chapter, we share some tools that can help students of teaching make sense of their field-based experiences and develop as professionals through "thinking like a teacher" exercises. Regular feedback from the partner

Table 8.1 Solutions to Field Experience Challenges

Challenge	Possible solutions
Need for quality field sites	• Video and written cases of exemplary teachers • Placements in informal settings (e.g., science museums, after school programs) • Partnerships with teachers and schools
Time	• Use cases and informal settings • Block out a few field visits in the semester calendar • Build into the university schedule
Communication	• Regular website and email communications • Written expectations for students, shared with partner teachers • School building liaison to communicate with partner teachers • Regular meetings among elementary education faculty
Assessment	• Clear expectations communicated to students, partner teachers, and supervisors • Non-threatening assessment tools • Focus on learning, not expertise
Learning from experience	• Team work • Regular opportunities for team and individual reflection (thinking like a teacher) • Regular feedback from partner teacher and university supervisor

teacher and university supervisor is another source of information for reflection that can reinforce learning from experience.

> **It's Your Turn**
>
> Select the challenge you think will be greatest for you in designing and implementing a field experience. Brainstorm some ways to address the challenge. Select the solution(s) that you think are most viable for your context. Enlist the help of the individuals you need to ensure your solution works.

Various Models of Science Methods Field Experiences

In our collective experience teaching elementary science methods, we have experienced and invented a variety of field-based opportunities for our students. In this section, we present several examples of field-based approaches, derived from our own experiences and those of our colleagues. These examples differ in the degree to which they address the challenges mentioned above, and in the potential to achieve various benefits. Their viability depends on the context of the methods course and the relationship that has been established between the university and the schools. However, we can attest that we have witnessed each of these ideas implemented with some degree of success in elementary science methods instruction.

Example 1: The Virtual Field Experience

The virtual field experience addresses some logistical challenges to field experiences by avoiding school-based placements altogether. Microteaching (Yeany, 1977, 1978) is the most typical example. Science methods students plan and teach lessons to each other, taking the roles of children and their teachers. In light of situated learning theory, we question if the learning of future teachers is jeopardized by placing them in an apprenticeship context so unlike what they will encounter in their future teaching. Nonetheless, one advantage of this approach is that it allows prospective teachers to simultaneously view a lesson from the perspective of learner and teacher. Another way for methods students to view science learning from multiple perspectives is to participate in a science lesson as learners, and then examine actual student work from the same lesson. Comparing their thinking with elementary students' allows prospective teachers to connect to elementary students as learners and to take on a teacher perspective as well.

Other types of virtual field experiences have been expedited via various kinds of distance technology, for example, classrooms linked via video to university courses (Boone, 2001) and email and web-based communities of elementary students and preservice teachers (Abell & Julyan, 1994). Another type of science methods virtual field experience has been accomplished via case-based instruction (e.g., Tippins, Koballa, & Payne, 2002). Cases provide windows into classrooms that are discussed at a distance. In Abell's teaching of elementary methods, she created interactive videocases (Abell, Bryan, & Anderson, 1998; Abell, Cennamo, & Campbell, 1996; Abell et al., 1996; Cennamo, Abell, George, & Chung, 1996) to present examples of reform-minded practice to methods students. The virtual field experience has numerous advantages: (1) instructors can select examples of practice consistent with their beliefs; (2) instructors can help prospective teachers focus their observations on particular instances that illustrate specific learning goals or teaching strategies; and (3) instructors can avoid many of the logistical hassles associated with real time field experiences. The major disadvantage of most virtual field experiences is that students do not have opportunities to legitimately participate inside the teaching community; rather they watch from afar.

Example 2: The Extracurricular Field Experience

In this model, the science teacher educator takes advantage of, or creates, science experiences for children outside of school that serve as the field setting for preservice teachers. For example, we have worked with students in existing science clubs or after school day care settings, and in informal science education settings such as parks, zoos, and museums, where the curriculum is open for new ideas and the need for extra adults is great (Hanuscin, 2004; Wissehr & Hanuscin, 2008). We also know science educators who have invented their own after school experiences, Saturday science programs, or summer camps as places for teaching interns to apply what they have learned in the real world. In Hanuscin's teaching of science methods, she was confronted with the challenge of teaching the course in the summer, when schools were not in session, and thus no field experience was possible. To address this, she engaged the class in creating a week-long science experience for elementary-age students (see Hanuscin & Musikul, 2007). The collaborative effort of designing, enacting, and reflecting on instruction proved beneficial in terms of encouraging risk-taking and exploration of unfamiliar pedagogies, such as inquiry. The experience was co-constructed by students and instructor, allowing them to develop a shared vision for effective science teaching.

An advantage of such settings over more traditional field experiences is that prospective teachers may have the opportunity to implement the same lesson with several different groups of students in a very short timeframe—versus the time it takes for classroom teachers to have this same opportunity. In doing so, prospective teachers get important feedback about the effectiveness of their teaching, as well as the opportunity to make adjustments and learn the consequences of their instructional decisions on student learning (Hanuscin, 2004). Prospective teachers enjoy the freedom allowed in these settings and the high interest levels of the students involved. A disadvantage is that these settings may fail to capture the classroom in all its complexity or may not represent the diversity of students in public schools.

Example 3: The Demonstration Lab

Another way to create experiences for preservice teachers involves creating a demonstration lab in a local school. Deborah Smith (2000) pioneered this model at the University of Delaware as a tool for the professional development of elementary teachers of science. In the demonstration lab, an experienced science educator models instruction with a group of students in a local elementary school, while the classroom teacher participates as an aide. Eventually, the roles are reversed and the classroom teacher becomes the instructor and the science educator the aide. This demonstration model has been used in the internationally acclaimed Reading Recovery program (Clay, 1993; Deford, Lyons, & Pinnell, 1991). In that program, a group of teachers watch a single teacher teach from "behind the glass," a two-way observation mirror behind which a discussion of the teaching and learning is led by an experienced facilitator. The demonstration model can also be applied in the preservice setting, where a group of interns observes an experienced teacher or science teacher educator and discusses her practice in a debriefing session. The advantages of the demonstration lab are that (1) models of best practice can be selected to observe, and (2) opportunities for reflection abound. Although there is only one placement to arrange, setting up that demonstration setting requires establishing trust with teachers, which is a long-term process. This model also faces the scheduling challenges of the add-on experience (see next section).

Example 4: The Add-On Field Experience

When faced with the prospect of having no field experience associated with their science methods course, many science teacher educators have attempted adding on a field experience, either bit by bit or as an integrated part of the course. In her teaching of science methods, Abell began adding field experiences slowly but deliberately to her campus-based methods course. As a start, she asked students to complete two assignments in the field. The first, interviewing science learners about their science ideas, was coordinated by the students themselves (see "Tools" section). The methods students found the children and scheduled the interviews outside of class time. Abell coordinated the second, a three-day teaching experience, by asking teachers to volunteer to accept a team of students into their classrooms. She spent hours fitting together schedules and teams as best as possible. Although this was a logistical challenge, she managed to grow the experience to include field placements for four to five course sections each semester, to partner with a large group of local teachers, and to involve students in several classroom-based projects across the semester. The add-on model, although not without challenges, can improve on the virtual experience or the demonstration lab by requiring preservice teachers to plan and enact instruction and reflect on firsthand evidence of student learning.

Example 5: The Partners Model

The add-on field experience can be strengthened, and logistical challenges reduced, by developing long term relationships with local teachers and schools, much as in the demonstration lab model. We have experienced two different, but reinforcing ways, of building such relationships. The first method involves working with local teachers in graduate courses and professional development projects. In these settings, teachers start to think differently about science teaching and learning, and begin to experiment with their own teaching. These teachers are ideal mentors for future teachers, because they are aware of current trends in the field and open to new ways of teaching science. Over time it is possible to build a cadre of science enthusiast teachers (Abell & Roth, 1992) who enjoy regular involvement with science methods field experiences. The second method requires an institutional commitment to building more formal partner schools arrangements. Collaborating with schools in terms of teacher preparation, professional development, and research makes sense given the community of learning we are trying to build around elementary teacher education. For universities that offer multiple sections of the science methods course each semester, partner schools can be identified for each section, with one teacher who serves as the building liaison with the methods course. This facilitates consistent communications concerning expectations and schedules. School/university partnerships are most successful when they exhibit common goals, mutual trust and respect (Abell, 2000), shared decision making, a commitment from teachers and administrators, and a manageable agenda with a clear focus (Robinson & Darling-Hammond, 1994). Although the work of building partnerships takes time and concerted effort, the potential outcomes are well worth it. The partners model, beyond relieving logistical and supervision challenges, builds a setting where teacher learning through apprenticeship with master teachers becomes the standard.

Example 6: The Field Experience Block

Because the future elementary teacher is often also enrolled in subject matter pedagogy courses in mathematics, literacy, and social studies at the same time as the science methods course, science teacher educators might explore ways to collaborate with university colleagues who teach these courses to create and schedule field experiences. The motivation is both logistical and pedagogical. Logistically, it can be more efficient to schedule students into a field experience block associated with one partner school, rather than have students enrolled in separate field experiences for each methods course. Pedagogically, students can benefit from observing elementary students and their teachers in more than one subject matter setting.

One challenge of the block model is to ensure that preservice teachers have the opportunity to observe and teach science during their field experience block. Another challenge is for instructors across disciplines to communicate and collaborate in designing ways for preservice teachers to achieve learning goals for several courses in the same field experience block. One solution to these challenges is to design a web-based portal where methods students, classroom teachers, and university supervisors discuss and reflect on problems of practice. Knowles et al. (1994) presented a theoretical framework for designing field experiences that require students of teaching to collect data, plan, enact instruction, and assess student learning in spiraling cycles throughout the teacher preparation program. This theoretical framework is useful for coordinating field experiences across an elementary teacher preparation program.

> It's Your Turn
>
> *What kind of field experience, if any, currently exists at your institution?*
> *What challenges are met by the design of the field experience? Which are not?*
> *Which of the above examples represents where you would like to go with your field experience in the coming year?*
> *Which represents your long-term goals for the field experience associated with your science methods course?*

Student Reflection on the Field Experience

Preservice teachers' prior beliefs about science teaching and learning shape their practices (see Chapters 3 and 4). To learn from the field experience, they need to confront these beliefs as they make sense of viable practices in elementary science. The examples will not lead to effective learning unless preservice teachers have explicit opportunities to reflect on and make sense of their experience. We have developed a number of reflection tools that help students build the authority of their experience in learning to teach science. With these tools, we encourage students to "think like a teacher" instead of like a student. In this final section of the chapter, we highlight several tools with examples from our own practice.

Making Sense of Introductory Observations

Because methods students have spent many years in classrooms as science learners in the "apprenticeship of observation" (Lortie, 1975), we believe that it is critical to their development to view classrooms from another angle, that of the teacher. We believe that this requires a structured approach to classroom observation. Before preservice teachers undertake classroom teaching in the field experience, we often ask them to observe the classroom in which they will teach. We have found that preservice teachers attend to the classroom setting better if we create a context for the observation and require a written reflection at the end. Because our students often focus on the teacher, not the students, and thus miss important connections between teaching and learning, we direct them to observe the students. We ask the future teachers to observe and write in response to a set of questions such as those in the "Tools" section of this chapter. Some of the foci for informal observation/reflection activities of learners include:

- A focus on an individual learner who is succeeding in science.
- A focus on an individual learner who is struggling in science.
- A focus on a group of students who are working as a team.
- A focus on student talk during a class discussion.
- A focus on which students participate in hands-on activities.

Whatever the observation context, we attempt to direct preservice teacher observations to a particular facet of classroom life. Providing questions that they must respond to in writing helps them pay close attention to learning and teaching. Oftentimes we tie the observation to a topic we have been reading about in class to help bring abstract concepts to life. For example, when we read about alternative conceptions, we ask the preservice teachers to pay attention to the science ideas that students voice. When we read about cooperative learning, we ask them to pay attention to what happens in small groups. Such activities provide a bridge between the field experience and the campus-based course.

Making Sense of Practice

In addition to observing others teach, future teachers need opportunities to plan, carry out, and assess instruction. We scaffold their performance as teachers by asking them to start small—teaching a short lesson to a small group of students, leading a short discussion, or doing a demonstration. Across the semester, their responsibilities might increase, until they are asked to plan and carry out a 3–5 day unit of instruction with their teaching team. During the act of teaching, beginning teachers are concerned about making it through the lesson without major problems. To prompt preservice teachers to pay closer attention to what they are doing and how students are responding, we develop written reflection assignments. We call these reflections "Thinking Like a Teacher," to contrast them with the activities we often do in class that ask the preservice teacher to think like an elementary student. The "Tools" section of this chapter includes a writing assignment given at the end of the semester of the methods course in conjunction with implementing a teaching unit in an elementary classroom. The assignment asks preservice teachers to evaluate the teaching and learning that occurred and provide evidence for their claims. The assignment also asks the teachers to think about how they might change their instruction in future enactments. Such assignments help students recognize that teachers are lifelong learners involved in a process of continuous quality improvement.

Studying One's Teaching

The ultimate reflection that preservice teachers can undertake often occurs at the end of a teacher education program, and into one's teacher career. This powerful form of reflection occurs when teachers have the opportunity to study their teaching through action research (Hubbard & Power, 2003). Some science educators include action research within the elementary methods course itself (van Zee, Lay, & Roberts, 2000). Methods course instructors can also implement shorter and more focused teacher-as-researcher assignments, such as interviewing elementary students about their conceptual understanding (see "Tools" section). Data from the interview project can inform the design and implementation of a unit of instruction in the methods course. Furthermore, techniques employed in a small teacher-as-researcher assignment can form the foundation for full scale teacher-directed action research projects later in the teacher preparation program. Although it is beyond the scope of this book to provide details for getting teachers involved in action research, these guidelines are available in other books (Hubbard & Power, 1999, 2003; Mills, 2002). We consider teacher research to be the ultimate vehicle for reflection on teaching.

> **It's Your Turn**
>
> Think about the form of your upcoming field experience. Now craft a written reflection assignment for some part of that field experience in which you ask the methods students to make some connection between a topic in the on-campus course and an observation or teaching assignment in the field experience. Make sure your reflection assignment prompts students to think deeply about elementary students' learning.

Conclusion: Design Principles for Science Methods Field Experiences

Based on our experiences designing and implementing various versions of field experiences associated with the elementary science methods course, we offer the following guiding principles for the design of elementary science field experiences (based on the backwards design model presented in Chapter 7).

1. Start with the learning goals. What kinds of learning outcomes do you expect from the field experience? Can you frame those goals in terms of situated vs. rule-bound learning?
2. Design ways to assess student learning. Students need guidance in learning from experience. We need to ask students to think about their teaching situationally, rather than asking them to apply abstract rules to generic situations. How will methods course students demonstrate their learning? What products will you require at what points in the field experience? How will you provide feedback? Grades?
3. Provide environments conducive to teacher learning. This includes classrooms where teachers are open to teaching in reform-minded ways, where teachers are prepared to coach newcomers into the field, and where the learning goals are agreed upon by all stakeholders.
4. Be collaborative and communicative. Successful teacher education is not owned by the university, but shared between the schools and the university. When we regard our school-based colleagues as team members in the field experience and develop shared expectations for the methods students, we prevent many problems from ever arising.
5. Make the field experience settings social. Teams of preservice and veteran teachers can work together to understand problems of practice. By building learning communities (see Putnam & Borko, 2000) and make their functioning explicit for all participants, we contribute to the learning of all who are involved.

Field experiences, although challenging to establish and maintain, can be a value-added component of the elementary science methods course. Carefully selected placements allow students of teaching to observe science learning and teaching in safe settings where they can make some of their own teaching decisions. Carefully planned reflection assignments will help students of teaching learn from the authority of their experience.

> Carl and I met several times to plan our first science methods field experience. We decided to start small to ensure success. We enlisted the collaboration of 10 local teachers who had been part of a summer workshop about teaching electricity concepts in the elementary school. We decided that the methods students would come to these teachers' classrooms in teams of three, four times during the semester. The first time would be to observe the teacher teaching a science lesson and debrief with her about her thinking. The second time would be to interview the children about their electricity ideas. During the third and fourth visits, the preservice teachers would be to carry out an "Engage" and "Explore" activity in the electricity unit (see Chapter 6 and ATW 1). One of the partner teachers agreed to film her teaching of the "Explain" phase that we would then show in class to wrap up the field experience. We decided that, concurrently with the field experience, we would carry out ATW 5, related to learning about electricity. We are excited to put our plan into action. We know we will encounter some obstacles, but we think we have a solid plan that will lead to loads of learning for our methods students and for the children in the elementary classrooms.

Tools for Teaching Elementary Science Methods

Sample Field Experience Feedback Form

Date _____ Partner Teacher _____

Teaching Team _____

Lesson Topic _____

For each section below, please list the positive aspects of your observations and give suggestions for improvement to the preservice teachers.

I. Comments on the Science Learners (engagement; understanding; age-appropriateness; special needs).

II. Comments on Preparation and Classroom Management (e.g., science content knowledge; materials management, transitions, cooperative learning).

III. Comments on Teaching Behaviors (questioning, responding, use of inquiry-based strategies).

Reflection on Initial Observations

After the observation: write a one to two page paper describing the learner who you observed.

1. What surprised you about your observation? What did you expect?
2. What did you learn about students at this age and their science learning?
3. How will you need to structure your small group lesson plan to accommodate the characteristics of learners of this age?

Reflecting on Teaching a Unit: Thinking Like a Teacher

The purpose of these reflections is for you individually to critically analyze your planning and teaching, and your students' learning. It will require completion of three informal reflections and one structured analysis. This is to be done individually, not as part of your teaching team.

Reflection #1

After your first day of unit teaching, write a one-page reflection. Assess how your teaching went today. Describe some of the *specific* interactions (quotes if you can remember them) with the children that help you to evaluate the success of achieving the content and process goals of your lesson. You should include thoughtful answers to the following questions:

- What were some of the interesting questions that the students asked?
- How did you address these questions?
- Was there something unexpected that happened? Explain.
- What were some of the students' reactions to your lessons?
- How were these reactions similar and different to what you would expect at this age level?
- What evidence do you have that students learned or did not learn?
- Was there a critical incident (something memorable) that occurred during your teaching?
- What science did you learn through your teaching?
- What are some of the improvements that you will make for the next lesson?

Reflection #2

After your second day of unit teaching, answer the same questions as in #1 (see above). However, this time focus your reflection on one or two students in your class with special needs or abilities. Discuss what adaptations your team will make for next class to better address his/her needs.

Reflection #3

Focus your reflection after the third day of unit teaching on a particular classroom problem that has arisen during your teaching (e.g., about a particular student, about a management issue, about how to help students learn a particular concept). Describe the problem by telling a story from your teaching over the past three lessons. Brainstorm possible solutions to the problem. Describe the solution you think would work the best and why. Try to implement this solution to some degree on your last day of teaching.

Final Reflective Analysis

Your written analysis should be two to three typed pages. It should include:

a. A brief description of your lesson's goals for student learning (two to three sentences).
b. A detailed analysis of your assessment of student learning in light of your goals (two to three paragraphs). Briefly discuss the assessment task used (one to two sentences). Expand upon the children's performance. Did they meet your expectations? Why or why not? Were there patterns in the children's responses? How do you explain them? Give evidence for your analysis by quoting children's written responses from the assessments. Attach examples of student writing and drawing as evidence of your assessment of their learning and refer to these as specific evidence of student learning. Also refer to course readings on interpreting children's work.
c. An analysis of your teaching (two to three paragraphs). Think back upon your unit teaching. Refer to your reflections from previous days. Don't reiterate your unit plan. What were some of the strengths and weaknesses of your teaching? Did the assessments that you used throughout your lesson plan (brainstorming, journal writing, drawing, observations, oral responses and discussion, self-assessments, etc.) foreshadow the results of the final assessment? Why or why not? Explain using specific examples from your 3 days of teaching (this can include student responses, either written or oral). Refer to and cite course readings on teaching for understanding.
d. A plan for the future (two to three paragraphs). For our teaching to be effective, we must provide students with opportunities to construct, apply, and restate concepts. Keeping in mind what you know about these students' learning and your teaching, where would you go with this lesson if you had another 2 weeks to teach? How would you adjust your teaching and assessment practices to increase the effectiveness of your teaching? Move beyond statements such as: I didn't have enough time, materials, knowledge of students, etc. Provide specific ideas for changing your teaching.

Interview Assignment as a Form of Teacher Research

Purpose

The purpose of this assignment is for you to gain experience with a potent technique for assessing students in science—the individual interview. In this activity you will devise, administer, and analyze an interview with elementary students on a particular science concept.

Procedure

1. Choose a science concept. Make sure you have an accurate and complete scientific understanding of the concept.
2. Develop an interview about *events* or *instances* (see examples on course website) that will help you ascertain student ideas about this concept. Course readings will provide helpful ideas for questions and tasks.
3. Practice giving the interview to friends, relatives, classmates. The article called "Finding out What Children Think" by Bell, Osborne and Tasker (1985), will be helpful in conducting interviews.
4. Interview Day. Bring to school site: tape recorder, tape, extra batteries, drawing materials (paper and writing utensils), copy of interview protocol, and props. Tape record the interview.
5. Transcribe the interview verbatim. Be sure to indicate child's age (first names only). Format might be as follows:

    I: (interviewer) "What you actually said."
    C: (child's response) "Use exact quotes."

6. Next read the transcript and highlight in two colors to indicate: (1) good interviewing practices; and (2) improvement needed in interviewing practices. Include a key.
7. Write and type a short (three to four page) analysis of the interview that includes:
    a. An analysis of your skills as an interviewer (see Figure A-1 Checklist for Interviewers, following the Bell, Osborne and Tasker article. Give specific examples from your interviews).
    b. Analysis of children's responses:
        - Discuss the scientifically appropriate responses to interview questions.
        - Compare and contrast the students' ideas with those of scientists. (Use student quotes and drawings to support your ideas.)
        - Compare and contrast the students' ideas with those in the research (see course readings).
        - Discuss the implications to teaching and learning, i.e., how this information might affect the design and teaching of your unit.

8. Bring to class: two copies transcripts and analysis—one to hand in and one to use with your group in class. In class you will compare your findings with those of the other students in your group and synthesize what you learned about the students and how that will affect your teaching and their learning. Your group will make a presentation (see below) to the class of this synthesis.

Group Presentation

Your group will compare results from your interview and prepare a slide presentation using PowerPoint or other way to prepare slides. Your presentation should contain the following:

- Title slide: gives title and students who worked on the presentation.
- Focus of the interview: what you were trying to find out about the students' understanding. Include a representation of the science concept(s) addressed (e.g., concept map, diagram, outline).
- Scientifically correct view of your concept.
- Interview procedures: how you conducted the interview. Numbers and ages of students interviewed. Questions asked.
- Results: patterns that you found among the children interviewed by your group. Can be in the form of a table. Include student drawings and quotes as evidence (first names only).
- Comparisons to what the research says (e.g., Driver book).
- Teaching implications that will affect the way you teach your unit.
- Closing slide.

Evaluation

This assignment will be graded in three parts—the protocol itself (thorough, accurate), your analysis of the interview (all components addressed, insightful), and your group's synthesis presentation (visually appealing, thoughtful and complete, well organized).

References

Abell, S. K. (2000). From professor to colleague: Creating a professional identity as collaborator in elementary science. *Journal of Research in Science Teaching, 37*, 548–562.

Abell, S. K. (2006). Challenges and opportunities for field experiences in elementary science teacher preparation. In K. Appleton (Ed.), *Elementary science teacher education: Contemporary issues and practice* (pp. 73–89). Mahwah, NJ: Erlbaum.

Abell, S. K., Bryan, L. A., & Anderson, M. A. (1998). Investigating preservice elementary science teacher reflective thinking using integrated media case-based instruction in elementary science teacher preparation. *Science Education, 82,* 491–510.

Abell, S. K., Cennamo, K. S., Anderson, M. A., Bryan, L. A., Campbell, L. M., & Hug, J. W. (1996). Integrated media classroom cases in elementary science teacher education. *Journal of Computers in Mathematics and Science Teaching, 15,* 137–151.

Abell, S. K., Cennamo, K. S., & Campbell, L. M. (1996). Interactive video cases developed for elementary science methods courses. *Tech Trends, 41*(3), 20–23.

Abell, S. K., & Julyan, C. (1994). Educational journeys in college classrooms. In C. L. Julyan & M. S. Wiske (Eds.), *Learning along electronic paths: Journeys with the NGS Kids Network* (pp. 236–243). Cambridge, MA: Technical Education Research Center.

Abell, S. K., & Roth, M. (1992). Constraints to teaching elementary science: A case study of a science enthusiast student teacher. *Science Education, 76,* 581–595.

Bell, B., Osborne, R., & Tasker, R. (1985). Finding our what children think. In R. Osborne & P. Freyberg (Eds.), *Learning in science: The implications of children's science* (pp. 151–165). Portsmouth, NH: Heinemann.

Boone, W. J. (2001). A qualitative evaluation of utilizing quad screen technology for elementary and middle school teacher preparation. *Journal of Technology and Teacher Education, 9,* 129–146.

Brown, J. S., Collins, A., & Duguid, P. (1989). Situated cognition and the culture of learning. *Educational Researcher, 18*(1), 32–42.

Cennamo, K. S., Abell, S. K., George, E. J., & Chung, M. (1996). The development of integrated media cases for use in elementary science teacher education. *Journal of Technology and Teacher Education, 4,* 19–36.

Clay, M. M. (1993). *Reading recovery: A guidebook for teachers in training.* Portsmouth, NH: Heinemann.

Clift, R. T., & Brady, P. (2005). Research on methods courses and field experiences. In M. Cochran-Smith & K. M. Zeichner (Eds.), *Studying teacher education: The report of the AERA panel on research and teacher education* (pp. 309–424). Mahwah, NJ: Erlbaum.

Deford, D. E., Lyons, C. A., & Pinnell, G. S. (1991). *Bridges to literacy: Learning from reading recovery.* Portsmouth, NH: Heinemann.

Druckman, D., & Bjork, R. A. (1994). *Learning, remembering, believing: Enhancing human performance.* Washington, DC: National Academy Press.

Hanuscin, D. (2004). A workshop approach: Instructional strategies for working within the constraints of field experiences in elementary science. *Journal of Elementary Science Education, 16*(1), 1–8.

Hanuscin, D., & Musikul, K. (2007). School's IN for summer: An innovative field experience for elementary science methods. *Journal of Elementary Science Education, 19*(1), 57–68.

Hubbard, R. S., & Power, B. M. (1999). *Living the questions: A guide for teacher-researchers.* York, ME: Stenhouse Publishers.

Hubbard, R. S., & Power, B. M. (2003). *The art of classroom inquiry: A handbook for teacher-researchers* (rev. ed.). Portsmouth, NH: Heinemann.

Knowles, J. G., Cole, A. L., with Presswood, C. S. (1994). *Through preservice teachers' eyes: Exploring field experiences through narrative and inquiry.* New York: Merrill.

Lave, J. (1988). *Cognition in practice: Mind, mathematics and culture in everyday life*. Cambridge, United Kingdom: Cambridge University Press.

Lave, J., & Wenger, E. (1991). *Situated learning: Legitimate peripheral participation*. Cambridge, United Kingdom: Cambridge University Press.

Lortie, D. C. (1975). *Schoolteacher: A sociological study*. Chicago: University of Chicago Press.

McIntyre, D. J., Byrd, D. M., & Foxx, S. M. (1996). Field and laboratory experiences. In J. Sikula (Ed.), *Handbook of research on teacher education* (2nd ed., pp. 171–193). New York: Macmillan.

Mills, G. (2002). *Action research: A guide for the teacher researcher* (2nd ed.). Upper Saddle River, NJ: Prentice Hall.

Munby, H., & Russell, T. (1994). The authority of experience in learning to teach: Messages from a physics methods class. *Journal of Teacher Education, 45*, 86–95.

Putnam, R., & Borko, H. (2000). What do new views of knowledge and thinking have to say about research on teacher learning? *Educational Researcher, 29*(1), 4–15.

Robinson, S. R., & Darling-Hammond, L. (1994). Change for collaboration and collaboration for change: Transforming teaching through school–university partnerships. In L. Darling-Hammond (Ed.), *Professional development schools: Schools for developing a profession* (pp. 203–219). New York: Teachers College Press.

Rogoff, B. (1990). *Apprenticeship in thinking: Cognitive development in social context*. Oxford, United Kingdom: Oxford University Press.

Schön, D. A. (1983). *The reflective practitioner*. New York: Basic Books.

Schön, D. A. (1987). *Educating the reflective practitioner: Toward a new design for teaching and learning in the professions*. San Francisco: Jossey-Bass.

Schön, D. A. (Ed.). (1991). *The reflective turn*. New York: Teachers College Press.

Shulman, L. S. (1986). Those who understand: Knowledge growth in teaching. *Educational Researcher, 15*(2), 4–14.

Smith, D. C. (2000). Content and pedagogical content knowledge for elementary science teacher educators: Knowing our students. *Journal of Science Teacher Education, 11*, 27–46.

Tippins, D. J., Koballa, T. R. Jr., & Payne, B. D. (Eds.). (2002). *Learning from cases: Unraveling the complexities of elementary science teaching*. Boston: Allyn and Bacon.

van Zee, E., Lay, D., & Roberts, D. (2000, April). *Fostering collaborative research by prospective and practicing elementary and middle school teachers*. Paper presented at the annual meeting of the American Education Research Association, New Orleans, LA.

Wilson, S. M., Floden, R. E., & Ferrini-Mundy, J. (2002). Teacher preparation research: An insider's view from the outside. *Journal of Teacher Education, 53*, 19–204.

Wissehr, C., & Hanuscin, D. (2008, January). *Science museums and specialized content courses for prospective elementary teachers: Implications for learning to teach science*. Paper presented at the annual meeting of the Association for Science Teacher Education, St. Louis, MO.

Yeany, R. H. (1977). The effects of model viewing with systemic strategy analysis on the science teaching styles of preservice teachers. *Journal of Research in Science Teaching, 14,* 209–222.

Yeany, R. H. (1978). Effects of microteaching with videotaping and strategy analysis on the teaching strategies of preservice science teachers. *Science Education, 62,* 203–207.

Part II

Activities that Work for the Elementary Science Methods Course

Introduction

The first part of this book explains principles and processes for teaching the elementary science methods course. They are presented in the form of propositional knowledge (Shulman, 1986) for teaching teachers. That is, we presented the research-based findings about science teacher education in the form of propositions. As a whole, these propositions form a foundation of what a science educator needs to know and be able to do in the context of the methods course. Although knowledge in the form of propositions is powerful, it tends to be simplified and decontextualized. The elementary science methods course itself is neither. We believe that methods instructors also need knowledge in a form that is contextualized to help build their PCK for teaching future science teachers. Case knowledge (Shulman) is represented by context-rich instances of practice that embody knowledge for teaching. Cases typically are narrative in nature, telling the story of practice through the eyes of the practitioner. In this section of *Designing and Teaching the Elementary Science Methods Course*, we present cases of teaching in the elementary science methods course in the form of Activities that Work (Appleton, 2006). The cases are told by a narrator—one of us—to give you a window into elementary science methods courses for a short time period.

Activities that Work

An interesting question to ponder is: how does a beginning teacher who has little PCK for science teaching begin to teach the subject? The same question can be asked about the beginning elementary science methods instructor. Of course, the beginning methods instructor has experienced an apprenticeship of observation, watching science teaching in action (Lortie, 1975), throughout his or her years as a science student. Many beginning methods instructors also have taken part in teacher preparation programs as part of their apprenticeship of observation. These experiences are limited, however, because they are based in the student perspective. Most methods instructors have also been teachers of science at some level. However, knowledge for teaching science is only a part of the knowledge base for teaching future teachers. To develop PCK for teaching teachers, we believe methods instructors need direct experience teaching methods courses, ideally with the guidance of more experienced others (see Abell, Park Rogers, Hanuscin, Gagnon, & Lee, 2009), who have a repertoire of knowledge for teaching teachers that they are usually happy to share with a beginner. However, because many beginning methods instructors do not have the opportunity to teach alongside an expert before taking the reins in their own methods course, we developed a set of cases for teaching elementary science methods, based on our own experience. We call these cases Activities that Work.

Appleton developed the notion of Activities that Work (2003, 2006) to help capture the PCK for teaching elementary science held by veteran elementary teachers. At their simplest, Activities that Work are brief descriptions of activities that engage elementary students in learning science that have stood the test of classroom implementation. Activities that Work may include details about management, instructions, explanations, questions to ask, and forms of assessment. We believe that the notion of Activities that Work is easily transferred to the context of teaching future teachers. However, the published materials available to elementary science methods instructors are limited to a small number of textbooks written directly for the methods student. These texts seldom contain explicit material for the methods course instructor. For these reasons, we include in this section a number of Activities that Work that can help a methods instructor construct PCK for teaching science teachers.

We consider these Activities that Work in the methods course to fill several functions:

1. They are practical for the context of a typical methods course in terms of the timeframe, available equipment, etc.
2. They help teach important science and nature of science concepts that prospective elementary teachers might not understand but that they likely will be required to teach.
3. They include a focus on specific pedagogical principles of science teaching and learning that are critical for the future teacher of elementary science.
4. They require methods students to alternate between the roles of science learner and science teacher.
5. They model best practices in science teaching for elementary teachers, often by embedding Activities that Work for elementary science learners within the methods course Activity.

Moreover, these Activities that Work are useful in connecting course learning activities into a coherent framework, because they allow a simultaneous focus on multiple goals.

Using Activities that Work to Teach Prospective Elementary Teachers

In Chapter 5, we presented a number of goals for the elementary science methods course. For this section, we selected a subset of these learning goals to describe through each Activity that Works. We purposefully designed activities for our methods courses to meet goals for methods student learning in three major areas simultaneously—science pedagogy, science content, and nature of science content. These goals helped to structure each Activity that Works.

First, each Activity that Works is designed to emphasize a pedagogical principle for teaching science in the elementary school, thereby helping prospective teachers establish a link between theory and practice. We model the principle in action by engaging our methods students in science activities appropriate for elementary students. After the science activity, we step back and debrief. Often we ask our students to think about the role of the instructor and the role of the student in the Activity. We might ask them to read articles about the pedagogical principle under study, watch videos of classrooms that display the principle in action, and discuss the value of the pedagogical ideas.

Second, each Activity that Works is designed to address specific science content. We select concepts aligned with national or state science teaching standards for the elementary grades. Over the semester, we address science concepts appropriate for different grade ranges (e.g., K-2, 4–6) and within different science disciplines (e.g., life sciences, physical sciences, earth and space sciences). As the research shows (see Chapter 4), many of our methods students struggle to understand these concepts and lack confidence in teaching them. Thus we believe that emphasizing science learning goals is a necessary component of the science methods course. To help our students learn the science concepts, we typically use a teaching model such as the learning cycle approach in which students explore a phenomenon and later work together with each other and the instructor to develop viable models and explanations. We ask students to be metacognitive about their learning and gauge their progress in understanding over time.

Third, each Activity that Works is designed to address knowledge about the nature of science. All instruction in the science and/or methods classroom implicitly conveys an image of science to learner. However, decades of research demonstrate that to teach students important ideas about the nature of science more effectively, teachers must use explicit versus implicit teaching strategies (Lederman, 2007). This does not mean that teachers

Table AI.1 Goals for Each "Activity that Works"

Activity	Pedagogical principles	Science topics (science discipline/grade level)	NOS concepts
1	The learning cycle	Magnets (Physical Science/K-3)	The role of observation and inference in science
2	The interactive approach	Sinking and floating (Physical Science/K-5)	How scientists judge the quality of explanations
3	Guided and open inquiry	Insect life cycles (Life Science/K-3)	Multiple types of scientific investigation
4	Eliciting student ideas	Human body (Life Science/K-4)	Scientists classify living things or parts by function
5	Use of models and analogies	Electrical circuits (Physical Science/4-6)	Scientists use analogies/models to explain phenomena that cannot be observed directly
6	Science classroom talk	Light and shadows (Physical Science/K-3)	Importance of models in science
7	Integration of science and language arts	The water cycle (Earth and Space Science/4-6)	Tentative nature of scientific ideas
8	Seamless assessment	Phases of the moon (Earth and Space Science/4-6)	Science involves human creativity and invention, not merely discovery

must be didactic or lecture to students; rather, just as with other content, teachers must elicit students' prior knowledge of the nature of science, provide opportunities to explore new ideas about the nature of science, and reflect on their developing understanding.

Thus an Activity that Works in the elementary science methods course is one that serves multiple goals, engages students in doing, talking, and writing about science and the nature of science, and involves reflecting on science teaching and learning. Table AI.1 displays each Activity that Works from this volume and the learning goals it addresses.

It's Your Turn

Now it's time for you to take the leap. According to Shulman (1986), your propositional and case knowledge for teaching prospective elementary science teachers will be enhanced by the strategic knowledge you develop as you use the Activities that Work in your own teaching. "Strategic knowledge comes into play as the teacher [teacher educator] confronts particular situations or problems, whether theoretical, practical, or moral, where principles collide and no simple solution is possible" (pp. 12–13). We recommend the following process for building your strategic knowledge for teaching science teachers:

- Use Table AI.1 to help you find an Activity that Works which fits with the learning goals of your elementary science methods course.
- Read our account of the Activity and think about how to apply it in your setting.
- Try it out, modify it, embellish it.
- Think about what worked for you and your students, and what you would like to change the next time you use the Activity.

Before you know it, your PCK for teaching the elementary science methods course will include your own Activities that Work to guide your teaching of future science teachers.

References

Abell, S. K., Park Rogers, M. A., Hanuscin, D., Gagnon, M. J., & Lee, M. H. (2009). Preparing the next generation of science teacher educators: A model for developing PCK for teaching science teachers. *Journal of Science Teacher Education, 20*, 77–93.

Appleton, K. (2003, July). Pathways in professional development in primary science: Extending science PCK. Paper presented at the annual conference of the Australasian Science Education Research Association, Melbourne, Australia.

Appleton, K. (2006). Science pedagogical content knowledge and elementary school teachers. In K. Appleton (Ed.), *Elementary science teacher education:*

International perspectives on contemporary issues and practice (pp. 31–54). Mahwah, NJ: Lawrence Erlbaum in association with the Association for Science Teacher Education.

Lederman, N. G. (2007). Nature of science: Past, present, and future. In S. K. Abell & N. G. Lederman (Eds.), *Handbook of research on science education* (pp. 831–880). Mahwah, NJ: Lawrence Erlbaum.

Lortie, D. C. (1975). *Schoolteacher: A sociological study*. Chicago: University of Chicago Press.

Shulman, L. (1986). Those who understand: Knowledge growth in teaching. *Educational Researcher, 15*(2), 4–14.

Activities that Work I

Learning about the 5E Learning Cycle
Magnetism

Introduction

As discussed in Chapter 6, the learning cycle (Bybee, 1997) has been embraced as a teaching approach consistent with the goals of teaching science as inquiry. Research has shown that instruction that follows the learning cycle can result in greater achievement in science, better retention of concepts, improved attitudes toward science and science learning, improved reasoning ability, and superior process skills than would be the case with traditional instructional approaches. Science teacher educators have been disappointed to find, however, that preservice teachers may fail to grasp this model, even after extensive instruction. As emphasized in Chapter 4, this model often differs significantly from the kind of instruction prospective teachers may have received in their own K-12 science experiences. For example, many of the prospective teachers in our methods courses have experienced "activitymania" (Moscovici & Nelson, 1998), in which their teachers presented a series of disconnected activities that quickly moved from one concept to the next. Many believe that teaching more than one activity related to a single concept is redundant. As a result, we find methods students often fail to grasp the manner in which conceptual development occurs through multiple learning experiences that build upon one another.

In addition to lack of understanding of the learning cycle, prospective teachers' perceptions of their roles and beliefs about teaching and learning can serve as barriers to their implementation of this approach. Many prospective teachers enter their science methods courses following high school and college experiences that reflected a form of science instruction where teaching was telling and learning was listening. Providing opportunities for methods students to experience a learning cycle approach as learners can be critical to their understanding of the learning cycle. Also necessary, however, is the opportunity to plan instruction using the learning cycle as teachers. As science teacher educators, we have found that the learning cycle model itself can provide a venue for sequencing learning experiences to help preservice teachers understand and apply this approach (Hanuscin & Lee, 2008).

Activity Overview

In this "Activity that Works," we describe a lesson sequence in which we use the learning cycle as a model for our own instruction to assist elementary science methods students in developing their understanding of the learning cycle and how it applies to their own teaching. The exemplar lesson embedded within this learning cycle focuses on teaching the concept of magnetism to early elementary learners (K-3), and targets understanding aspects of the nature of science such as the reliance on both observation and inference to develop models that explain phenomena.

The learning goals for the learning cycle and magnetism Activity that Works include helping methods students build understanding of science teaching, science concepts, and the nature of science:

Pedagogical Knowledge Objectives. Understand how multiple experiences can help build deeper understanding of the concepts (vs. being redundant); be able to select and sequence activities appropriate to each phase of the learning cycle.

Science Content Knowledge Objectives. Understand that magnets attract some metals (e.g., iron, nickel, and cobalt) but not others; explain how magnets interact with each other (attract and repel).

Nature of Science Objectives. Understand the role of observation and inference in explanation; be able to identify what can be directly observed in regard to magnets and what cannot.

To implement the learning cycle and magnetism Activity in your methods course, you will need to consider the following:

- *Timeline*:
  - Engage: 30–45 minutes.
  - Explore: 2–3 hours spread over two class sessions with students working outside of class on their inventions.
  - Explain: 1 hour.
  - Elaborate: 1 hour out-of-class time if reflection paper assigned.
  - Evaluate: allow in-class time for peer-review and final presentations.
- *Materials*:
  - Engage: card sort sets of activities for learning cycles (see "Tools" section).
  - Explore: assorted magnets and materials (including both ferromagnetic and non-ferromagnetic metals, such as Canadian and U.S. coins); found objects with which to create magnetic inventions (students may bring from home).
  - Explain: chart paper or white boards.
  - Elaborate: preservice teachers' original card sort responses.
  - Evaluate: access to relevant Standards documents, resources, etc.

Table A1.1 Overview of the Methods Course Activities

Engage	Prospective teachers participate in a card sort activity to select and sequence learning activities. Each small group is given a set of activities related to a single science concept.
Explore	Prospective teachers participate, as learners, in a model learning cycle lesson on magnetism.
Explain	Prospective teachers debrief their experience with the learning cycle, creating a chart that outlines the activities of teacher and student during each phase of instruction.
Elaborate	Prospective teachers revisit their ideas from the card sort activity, and reflect on the changes in their thinking.
Evaluate	Prospective teachers design their own learning cycle lesson sequences related to a concept appropriate to the grade level of the class in which they are placed for their field experience.

Throughout the sections that follow, we describe our methods course activities for each of the five phases of our instruction about the learning cycle. While we present it as an "Activity that Works" for teacher educators, we also note that it simultaneously models an embedded "Activity that Works" for elementary teachers. Table A1.1 provides an overview of the lesson.

Activity

Engage Phase

Consistent with the intent of the engage phase of the learning cycle, our initial activity is intended to elicit methods students' prior knowledge and beliefs about teaching. We developed a card sort activity (see "Tools" section) in which methods students select and sequence learning experiences into learning cycles. We begin by using examples of learning cycle lessons from our own teaching experiences as well as those described in the literature (see for example, Abell & Volkmann, 2006). To these, we add alternative activities that could be utilized within that learning cycle, as well as "distractors"—didactic and "cookbook" activities that would be considered either antithetical to the learning cycle approach and/or only loosely connected to the conceptual storyline of that learning cycle. Each set of cards thus included 8–10 activities relating to the concept and ranging from highly teacher-centered to highly student-centered activities.

Working in small groups, our methods students are asked to reach consensus on the selection of five activities they feel best relate to the concept(s) or "fit" together. Next, we ask groups to arrange the activities they selected into what they think would be an appropriate sequence for instruction. We ask each group to provide a rationale for their selection of each activity, as well as its place in the instructional sequence. We encounter common orientations toward teaching and learning science (see Chapter 3) in this phase of instruction that include teacher-centered and didactic instruction that places the teacher in the role of dispenser of knowledge. For example, methods students often select activities that involve presenting vocabulary before students explore. Furthermore, in our experience, students tend to draw upon activities similar to those they have experienced as learners, and shy away from selecting unfamiliar activities. Their own content knowledge also plays a significant role in their decisions about which activities to choose to best develop the concept(s). For example, students often fail to see the value of particular activities in addressing children's potential misconceptions, and exclude those from their selection as a result. We tell the methods students that we are going to experience a science lesson as learners, and after that revisit their ideas as teachers.

Explore Phase

Because the learning cycle differs so greatly from the type of science instruction many of our students experienced as learners, we provide them with firsthand experiences learning science content through this approach. Students participate as learners in a learning cycle lesson that focuses on content appropriate to the elementary classroom. This provides an opportunity for prospective teachers to experience learning through the learning cycle, as well as obtain evidence of the model's effectiveness in developing their own understanding of the content. Our students are often surprised to find that they have misconceptions about the science concepts they will be expected to teach elementary students. By working through their misconceptions, they are able to understand how the lesson might impact their own students in a similar manner. The learning cycle outlined in Table A1.2 in which our methods teachers participate, is designed to confront the misconception that all metals are attracted to magnets. After developing a better understanding of how magnets interact with each other and other materials, they apply their ideas to the design of an invention that utilizes magnets to solve a problem.

Explain Phase

The explanation phase is critical for sense-making following the initial stages of the Activity. In this phase of instruction, we focus on clarifying the purpose of each phase of the learning cycle, based on the model lesson

Table A1.2 Model Learning Cycle Lesson on Magnets

Engage	Students brainstorm, in small groups, a list of items they use everyday that involve magnets. The teacher records students' ideas in a whole class discussion, organizing the information in a three-column chart (item/product, what it does/problem it solves, how magnets are used). The teacher provides a transition to the next phase by highlighting any uncertainty students had about how magnets were used and suggesting that they explore magnets further.
Explore	Students are provided a box of objects (including both ferromagnetic and nonferromagnetic metals), and are asked to sort them based on how they think they will interact with a magnet. Students then test their ideas, noting any discrepancies. The teacher next provides other magnets and asks students to explore how magnets interact with each other.
Explain	Students share their observations about how magnets interact with the objects; the teacher introduces the terms *attract*, *repel*, and *ferromagnetic*. Students then generalize the kinds of objects attracted by a magnet (some metals, but not others). The teacher indicates that only iron, nickel, and cobalt are attracted to a magnet, and asks students what they can infer about the metal objects that were attracted by the magnet versus those that were not attracted by the magnet (e.g., U.S. nickels do not contain nickel, but Canadian coins must contain some iron, nickel or cobalt). The Explain phase concludes with a debriefing about observation and inference and their role in science; students are asked to brainstorm phenomena which can be directly observed and that scientific knowledge which is inferential in nature. For example, the magnetic field of a magnet cannot be seen; however, one can observe how objects interact with a magnet.
Elaborate	Students are asked to apply their understanding of magnetic interactions by designing an invention (or modify an existing one) that uses a magnet. The teacher guides students through the invention process (identify a problem, pose a solution, evaluate the solution, etc.), and references the inventions students brainstormed earlier. Students build and test their prototypes.
Evaluate	Students create a brochure for their product that explains what it does and how magnets are used to solve a problem. Students participate in an "Invention Convention" and communicate their design process and understandings of how magnets interact with each other and other materials.

on magnets in which the methods students just participated. Following the magnets learning cycle, the methods students work in groups to debrief their magnets experience by completing a two-column chart outlining the specific activities of the teacher and students in each phase of instruction (see "Tools" section). This reflection is intended to shift their perspective from that of learner to that of teacher. Through this process, we emphasize the role of the teacher in facilitating the learning experience and highlight strategies complementary to the learning cycle approach such as cooperative learning and productive questioning. Additionally, we focus on understanding the model further through professional readings about this approach, written from the perspectives of science educators and classroom teachers (e.g., Brown & Abell, 2007; Moscovici & Nelson, 1998). It is important to note this is the first time that preservice teachers are formally introduced to the vocabulary to describe each phase of the learning cycle (e.g., "Engage," "Explore"). We discuss this aspect of our instruction explicitly to illustrate the premise of exploration before concept introduction on which the learning cycle is based.

Furthermore, the Activity lends itself to a closer examination of science content standards, such as those found in the U.S. *National Science Education Standards* (*NSES*) (National Research Council (NRC), 1996). Working in groups, methods students identify the relevant standards (in this case, Standard B: Physical Science, Standard E: Science and Technology, and Standard G: History and Nature of Science). Prospective teachers often have a "laundry list" view of teaching standards; however, this examination lets them envision how a single lesson sequence may accomplish multiple goals.

Elaborate Phase

It is important for prospective teachers to self-assess and reflect upon their new understandings about this model of instruction, as well as their orientations to teaching and learning science. To accomplish this, we ask our methods students to revisit and critique their initial selection and sequencing of the activities from the card sort activity conducted in the engage phase of the lesson. In our experience, they often change not only the activities they originally selected from among the cards, but also the sequence in which they would use the activities. However, they also recognize that the same activity might be used in different phases of the 5E model depending on the purpose and way in which it is introduced to learners. For example, groups often recognize that while reading a children's literature book might serve to engage elementary students in considering their ideas initially, identifying accurate versus inaccurate information in the book could provide a summative assessment of what the children learned. Furthermore, prospective teachers are able to suggest modifications to the didactic and teacher-centered activities they initially selected that would make them more appropriate to the learning cycle model. Overall, the prospective teachers become more adept at identifying learning experiences that are central to developing the conceptual storyline.

Though the methods students complete this evaluation with their original group members and share their ideas through whole-class discussion, we also encourage students to reflect individually on their new understandings. As one of our students wrote,

> I really like the techniques we have learned in our class about the learning cycle. This technique has created a whole new perspective for me of what it means to teach science. In class during the sequencing activity our group was given a set dealing with electrical circuits. We disagreed over whether one part would be a good idea to teach in our classroom. "The teacher introduces vocabulary terms to students such as 'circuit,' 'conductor,' 'insulator,' and 'current.'" "Students create an illustrated dictionary in their lab notebooks, drawing pictures that convey the meaning of each of these terms based on their own observations and investigations." I thought this would be a good idea to use in the classroom. I think there is a time where a teacher needs to build knowledge by teaching students the proper terms of what they are seeing—and I learned this is the Explanation phase. This knowledge will help add meaning to other things they are seeing. I think that adding pictures at the bottom will help students apply the definitions to what they have witnessed in class.

As illustrated above, methods student reflections allow us another means to assess the change in their ideas about teaching and learning science, and the depth at which they understand the purposes of the different phases of the learning cycle approach.

Evaluate Phase

To apply their new understandings to a new context, prospective teachers plan their own learning cycle and develop a conceptual storyline based on a concept/big idea appropriate to the elementary classroom. Harking back to the initial card sort task, they begin by developing a collection of activities that relate to the focal concept. Just as in the initial card sort activity, they select and sequence five activities, based on how they combine to form a coherent conceptual storyline. Prospective teachers then prepare an outline of their lesson idea that includes a rationale for how each activity relates to the concept of focus and targets potential student misconceptions. They bring their outlines to class and work in small groups to provide feedback to one another on the selection and sequencing of their activities in terms of the learning cycle model. Following this round of peer review, they refine their outlines further and hand them in to the instructor for additional feedback.

Finally, the methods students articulate in writing their conceptual storyline for the lesson. Unlike a traditional lesson plan, this storyline provides a rich narrative describing each phase of instruction including specific details about their rationale for the overall lesson and specific choice of activities, what they and students will say and do, questions that will be utilized to facilitate discussion, how materials will be managed, what criteria they will use to evaluate student work, etc. Finished "lesson plans" are thus usually from 6–10 single-spaced pages in length. The purpose of this level of detail is to enable the instructor to make a valid assessment of preservice teachers' depth of understanding of the learning cycle and instructional decision making in the design of the lesson; however, preservice teachers consistently report that this is a valuable exercise that helps them realize just how much they must consider in designing effective instruction. We emphasize to our students that even in briefer versions of "lesson plans" they will write in the future, they will still need to think through their lessons in this same depth. As one student reflected,

> Before, I had simply googled lessons for other classes and handed in a one page lesson plan ... I didn't really think about *how* I'd teach the lesson. Writing out the learning cycle was really helpful! It helped me see how a variety of activities can fit together to help different learners.

Making the Activity Work for You

As a model for planning instruction, the learning cycle Activity that Works "can help teachers 'package' important instructional goals into a developing conceptual 'storyline' that accommodates both selection and sequencing of learning opportunities" (Ramsey, 1993, p. 1). In doing so, teachers can avoid the use of episodic and fragmented instructional activities or "activitymania." In the preceding sections, we outlined our own conceptual storyline to help prospective teachers understand the learning cycle model. Through applying the learning cycle model in our own instruction, we have found an effective means for teaching our methods students about this approach. In essence, we are practicing what we preach by modeling the same kind of instruction we expect them to use with elementary students. The lessons we developed for our learning cycle Activity that Works function together to help methods students develop a deep understanding of powerful ways to select and sequence learning activities for their own science instruction.

Such an approach could be easily modified using a different model learning cycle lesson for preservice teachers to experience in the Explore stage. For example, if you have taught a learning cycle lesson with elementary students around another topic, then sharing anecdotes from your work with children including examples of their talk and writing and drawing can be a powerful way to illustrate the teaching model in action. Alternately, if you have particular learning goals for your methods students related to another science content area, you could easily substitute a different model learning cycle for the one we have included here. In either case, embedding a science learning cycle within the science teaching learning cycle will be a powerful learning experience for your methods students.

Tools for Implementing the Activity

Card Sort Sets of Activities for Learning Cycles[1]

Set #1. The goal of this lesson sequence is to help students understand science as a process of inquiry, and to make connections between the work of scientists and their own classroom science activities. This sequence is part of a broader goal for helping students understand the nature of science.

- Students are asked to draw a picture of a scientist and explain what they think scientists do.
- Students design and conduct their own experiments to investigate plant growth. Their findings are shared in a mini-conference at family science night.
- Students, in pairs, read nonfiction books about the work of scientists, and then share their ideas about what they read in a whole-class discussion.
- Students make a Venn diagram comparing their own work in science class to the work that scientists do.
- The teacher invites a local biologist to visit the class as a guest speaker. Students ask this scientist questions about the work he/she does with corn plants.
- The class takes a field trip to a local farm to hear about ways farmers are trying to increase their crop yield.
- The class adopts a plot of land in their schoolyard and creates a garden to help beautify the school.
- Students read Chapter 5 of their science book, which discusses how plants grow. They compare their ideas in small groups, and discuss how the ideas they read relate to what they have been learning.

1. Reprinted from Hanuscin and Lee (2008) with permission of the Association for Science Teacher Education. Material excerpted from pages 61–66.

- Students bring in different food dishes from home, in which corn is a main ingredient. They discuss the importance of corn to different cultures.

Set #2. The goal of this lesson sequence is to help students understand how shadows are formed, and the factors that influence the size and shape of shadows. This sequence will prepare students for future lessons focusing on the behavior and properties of light.

- The teacher reads *Bear Shadow*, a story about a bear that attempts to escape a shadow that seems to be chasing him. Students are asked to think critically about what aspects of the story could be real or not in terms of Bear's shadow.
- As a class, students go on a "shadow hunt"—identifying different objects that make shadows, where shadows appear, and what light sources are present and their location. They draw pictures to record what they observe.
- In pairs, students make shadows of their own using flashlights and a variety of objects, provided by the teacher. Students record their ideas about what causes a shadow, as well as what causes a shadow to be different sizes or shapes.
- The teacher poses the following problem to students, and asks them to respond individually:

  > The local puppet show will soon be putting on a shadow play production of *The Three Bears*. Unfortunately, the three bear-shaped puppets they have are all the same size! How will they ever be able to make Momma, Poppa, and Baby Bear using the same sized puppets? Using what you know about shadows, explain how you think they might solve this problem.

- The teacher explains to students that the size of a shadow is dependent on the distance of the object from the light source, as well as the distance of the object from the surface on which its shadow falls. The teacher guides students in constructing diagrams to depict these two factors.
- Students draw pictures to illustrate ways to make shadows bigger and smaller. They share their pictures in small groups, and compare their ideas in a whole-class discussion.
- Students work in groups to design investigations to answer the question: What effect does the distance of the light source from the object have on the size of the shadow that is produced?
- Using a stuffed Teddie Bear, the teacher demonstrates different ways to change the size of a shadow, by projecting light from the overhead projector. She moves the Teddie Bear closer and farther from the

screen, and then moves the overhead projector closer and farther from Teddie. She then brings out a bigger stuffed Bear, and asks the class to predict what she could do to make the shadows of the two bears the same size.

Set #3. This lesson sequence is designed to help students understand how rocks are formed, and to appreciate the diversity of rocks that exist. It is part of a larger unit on the rock cycle.

- Students use Golden Guides™ (kid-friendly field guides) with information about different kinds of rocks—to identify samples of rocks provided by the teacher (or brought from home).
- Students are asked to bring in a rock from home. They share these in a circle, and the teacher closes the sharing session by asking students to write about what they think a rock is and where rocks come from.
- Students are provided a sample of rocks, which they sort based on characteristics they determine such as color, texture, and whether they float/sink. The teacher challenges students to think of criteria that would/would not be useful to classify rocks (e.g., two rocks may be the same type of rock but be different size and shape).
- The teacher teaches students how to sing "The Rock Cycle Song" (to the tune of "Row, Row, Row Your Boat") to help them remember that the three types of rocks are igneous, metamorphic, and sedimentary.
- Students make posters of the rock cycle, using their textbook diagram as a guide, and hang these posters in the classroom.
- Students, as a class, brainstorm a list of ways that people use rocks. Students generate examples ranging from pet rocks and landscaping, to building materials and the basis for sculptures in art.
- The class participates in a "rock exchange" with a classroom in another state—preparing a box with rocks found in their local area to send to them. Once they receive the box from their "rock pals" they compare the properties of rocks they receive to rocks found in their local area, and suggest reasons they might be the same and/or different.
- The class takes a field trip to the nearby state park, where a park ranger gives a talk about the local geology of the area and how it has changed throughout history.
- Students explore an interactive website that explains the various stages of the rock cycle, and how rocks are formed and reformed through this cycle.

Set #4. The goal of this learning cycle is to help students understand that each plant or animal has different structures that serve different functions in growth, survival, and reproduction.

- Students make a collage using cut-out images from magazines and newspapers that illustrate the diversity of a particular structure among a group of animals, and then writes a paragraph about their ideas. For example, one student creates a collage showing the many different kinds of feet that birds have, and then writes a paragraph that explains how having different types of feet might help birds do things to help them to survive.
- Students use different tools to represent the variety of beaks of different birds (e.g., pliers, a straw, tongs) and explore how much and what kind of "food" these birds might be able to eat in different habitats. The teacher has prepared the "habitats" in advance, and each contains a different variety of foods. Students conclude that their "birds" might not survive well in some habitats, because they would be unable to eat enough food.
- The teacher presents each group of students with a variety of bird feathers (from a single type of bird) and magnifying lenses. Students explore dropping and waving the feathers around, and develop detailed drawings in their science notebooks that illustrate the differences and similarities they observe between the different feather types. Students draw inferences about how having different types of feathers help a bird survive (e.g., down feathers are soft and help keep the bird warm, while flight feathers are rigid and help the bird push the air).
- Students look through Golden Guides™ (kid-friendly field guides) to learn about different species of birds found in their area. Parents are encouraged to help their child birdwatch in their backyard, and identify different species they see.
- Students compare birds found in their local area to birds found in different places around the world. Students use geographical resources to learn more about the environments in which the birds live, and then compare different adaptations the birds have to help them survive in those environments. Each group focuses on a specific pair of birds (one local, one from afar) to compare and share with the class.
- The teacher asks students to imagine what their life would be like if they had a beak instead of a mouth. During circle time, students go around and share something they think would be different, living with a beak.
- A guest speaker from the raptor rehabilitation program brings in several birds and discusses the different adaptations raptors have that help them survive by catching and eating their prey.
- Students read a chapter in their science book about animal and plant diversity. Afterwards, they discuss factors that affect the survival of different species.
- The class places two bird feeders outside of the classroom with very different kinds of foods. They keep track of the different species of birds that visit one feeder, versus the other.

Set #5. The goal of this lesson sequence is to help students develop an understanding of a simple circuit, and a model for the way electric current travels through a circuit. It is the first lesson in a unit that explores electrical circuits.

- Students, in groups, play the Operation Game. Then, they explain individually in writing how the game relates to what they know about simple circuits. They identify the path of the current as it travels through various parts of the game, as well as why the buzzer sounds and the patient's nose lights at some times, but not others.
- Students, working in pairs, design their own circuit quiz-boards by following instructions provided by the teacher, and filling in questions and answers of their own choosing. Some students provide words and definitions, while others list math problems and answers. Pairs trade quiz-boards with other groups and try to answer the questions correctly.
- Students are provided a battery, bulb, and wire, and are challenged to find ways to make the bulb light. They keep a list of "ways that work" as well as "ways that don't work," then look for patterns in their observations to develop "rules" for lighting the bulb. Students compare their rules in pairs, and negotiate any disagreement by retesting their configurations of battery, bulb, and wire to observe whether the bulb lights.
- Students are provided a flashlight, which they take apart to identify the circuitry found within. They create diagrams and written explanations of how a flashlight works, identifying the path of conductors through which the electric current travels, as well as the insulators through which current cannot travel.
- Students, in groups, design experiments to determine the effect that various factors (number of bulbs, number of batteries, length of wires, etc.) have on the brightness of the bulbs in a circuit. Students use the brightness of the bulbs as an indicator of the amount of current flowing through the circuit.
- The teacher introduces vocabulary terms to students such as "circuit," "conductor," "insulator," and "current." Students then create an illustrated dictionary in their lab notebooks, drawing pictures that convey the meaning of each of these terms based on their own observations and investigations.
- Students, in pairs, explore a multimedia CD that explains how current travels in a simple circuit. They answer a series of questions at the end of the module to test their understanding.
- Students are given a "mystery box" in which there may, or may not, be an electrical connection. (There may be batteries, bulbs, and wires in any configuration inside.) Two wires extend from the side of the box. Students are invited to test their ideas, and explain what they think is inside the box using evidence from their investigations.

Set #6. In this lesson sequence, students consider the question: "What causes sound?" The goal of this lesson sequence is to help students understand that "Sound is produced by vibrating objects." The mechanism of sound production is an important precursor to understanding other properties of sound such as "pitch" so it is the first lesson in a curriculum unit about sound.

- Students partner together to design and build an instrument. The whole class listens as each pair shares their instrument and presents how it produces sound.
- The students make sounds by using different materials that the teacher has placed at several stations around the classroom. They pluck rubberbands, blow into bottles of various shapes, and try out various percussion instruments while recording ideas and observations in their science notebooks.
- The teacher reads aloud the children's book *Very Quiet Cricket* by Eric Carle. Here is a brief synopsis of what the book is about. A cricket is born who cannot talk! A bigger cricket welcomes him to the world, then a locust, a cicada, and many other insects, but each time the tiny cricket rubs his wings together in vain: no sound emerges. In the end, however, he meets another quiet cricket, and manages to find his "voice." "And this time ... he chirped the most beautiful sound that she had ever heard."
- Students make and use "string telephones" out of tin-cans and string. Their recorded observations and ideas about what they think is happening to the sound become the focus of a class discussion.
- Students watch a video entitled "What is Sound?" and fill out a worksheet that the teacher created to go along with the video.
- Students are given the challenge to "Make the Sound Stop!" Each group has a buzzer and a limited amount of materials and time to figure out a way to stop the sound ... without turning off the buzzer, of course!
- Students are given a bag of materials: a tuning fork, small rubber hammer to strike the tuning fork, a jar of water, ping-pong ball. The teacher encourages students to find different ways to use the materials to make sound.
- Students write down three things that they know about sound on post-it notes. Groups of students compare and contrast their ideas with each another, noting that they don't all agree. This leads to a whole class discussion about students' ideas.

Set #7. The goal of this lesson sequence is to help students understand that almost all kinds of animals' food can be traced back to plants. This lesson is within a unit focused on the interdependence of organisms to each other and to their environment.

- The teacher shows a metal chain with links to students. She uses this as a model to represent what a food chain is. Students then use strips of paper, writing organism names and linking them to create their own "food chain." The teacher staples all of the food chains onto a bulletin board with a large sun in the center.
- Students choose a group of animals that interested them: pets, sea creatures, insects and spiders, etc. The teacher (with support from her school librarian) helps students find books and Internet resources so they can find out what their chosen animals eat. When students present their findings, the class constructs a Venn diagram: two overlapping circles labeled "animals that eat animals" and "animals that eat plants" to consider food resources.
- Students are given signs that have different plant and animal names and pictures on them. Students consider different relationships of who eats what, physically representing relationships by joining hands. At the conclusion, the teacher asks students to fill out an exit ticket:
  - Your friend's younger brother says: "Almost all kinds of animals can be traced back to plants." Do you agree or disagree? Why do you think so? Explain and then draw an example to support your answer.
- The teacher asks students to sort common food items (pictures, toys, or real) into three categories: Comes from a plant, Comes from an animal, or Both. Students work in small groups to discuss and sort the foods. Students write names of food items on a large chart drawn on the class chalkboard. The whole class then discusses whether plants and/or animals are needed as food.
- Students write responses to the following questions:
  - What would happen to an owl living in a forest if all of the mice could not find food? Why?
  - What would happen to the bees in an area if all the flowering plants died? Why?
- The teacher reviews the definitions of a carnivore, omnivore, and herbivore. Students look in old magazines to find pictures of each type of animal to place on a poster board.
- Students want to know what foods like marshmallows, candy, and mayonnaise are made of because the source isn't obvious to them. Students bring in food labels and work in groups to figure out if the ingredients can be traced to plants or not.

Sample Chart of Teacher/Student Activities during Learning Cycle

The learning cycle

Phase of instruction	Activities of the teacher	Activities of the students
Engage	Establish a context for study. Motivate students. Identify students' current science ideas and misconceptions. Figure out what students need to explore in the next phase.	Connect past and present learning experiences. Start thinking about concept to be explored. Get motivated and interested.
Explore	Provides a common set of experiences for students. Determine how students are processing in their conceptual understanding. Determine what students need explained in the next phase.	Clarify and test their ideas against new experiences. Compare their ideas with ideas of their peers and the teacher.
Explain	Provides opportunities for students to use previous experiences to begin making conceptual sense of prior explorations. Introduces formal language, scientific terms, and content information as needed. Determines what concepts need further instructional attention. Determines what elaborations will help scaffold learning in the next phase.	Demonstrate their current understandings. Develop explanations based on prior experiences. Use formal language, scientific terms, and content information to aid them in describing and explaining.
Elaborate	Provides opportunities to apply or extend the students' developing ideas through new activities. Assesses how students use formal representations of science knowledge (terms, formulas, diagrams). Determines what will be important to evaluate in the next phase.	Apply and transfer their knowledge and skills in new contexts. Relate past experience to current activities. Communicate their current ideas.
Evaluate	Assesses what students understand and can do at this point. Encourages students to be metacognitive. Determines what should occur in subsequent learning cycles.	Assess their own understandings as they solve problems. Be metacognitive about their learning.

Adapted from Abell and Volkmann, 2006.

More to Explore

Moyer, R. H., Hackett, S. A., & Everett, S. A. (2007). *Teaching science as investigations: Modeling inquiry through learning cycle lesson.* Columbus, OH: Prentice Hall.

Rubba, P. A. (1992). The learning cycle as a model for the design of science teacher preservice and inservice education. *Journal of Science Teacher Education, 3,* 97–101.

Settlage, J. (2000). Understanding the learning cycle: Influences on abilities to embrace the approach by preservice elementary school teachers. *Science Education, 84,* 43–50.

References

Abell, S. K., & Volkmann, M. J. (2006). *Seamless assessment in science.* Portsmouth, NH: Heinemann.

Brown, P. L., & Abell, S. K. (2007). Examining the learning cycle. *Science and Children, 44*(5), 58–59.

Bybee, R. W. (1997). *Achieving scientific literacy: From purposes to practices.* Portsmouth, NH: Heinemann.

Hanuscin, D., & Lee, M. H. (2008). Using the learning cycle as a model for teaching the learning cycle to preservice elementary teachers. *Journal of Elementary Science Education, 20*(2), 51–66.

Moscovici, H., & Nelson, T. (1998). Shifting from activitymania to inquiry. *Science and Children, 35*(4), 14–17, 40.

National Research Council (NRC). (1996). *National science education standards.* Washington, DC: National Academy Press.

Ramsey, J. (1993). Developing conceptual storylines with the learning cycle. *Journal of Elementary Science Education, 5*(2), 1–20.

Activities that Work 2

Interactive Approach
Floating and Sinking

Introduction

Chapter 6 concluded with some example teaching models that fit within the umbrella of inquiry. In the elementary classroom, these teaching models are intended to scaffold elementary student learning over a number of science lessons. However, prospective teachers tend to have considerable difficulty grasping the idea of sustained inquiry into one topic over several lessons. We believe this is for two main reasons. First, they seldom, if ever, have participated in a long-term science learning experience and thus may not understand that science learning takes time. Second, in field experiences (see Chapter 8), we seldom provide opportunities for methods students to observe the long-term development of students' ideas, or to plan and enact a sequence of lessons. It is therefore important for them to experience the teaching models that take place over several lessons in their science methods course. Ideally, experiences with a teaching model should be spread over several weeks. Sometimes, however, it is not possible for the teaching approach to be modeled exactly as it would be used in an elementary classroom due to time constraints in the methods course. Fortunately, as adult learners, prospective teachers can understand compressed versions of the teaching models that fit within one or two sessions.

There are key pedagogical ideas that should be emphasized while modeling the Interactive teaching approach: (1) determine what students already know about the topic, (2) clarify student learning goals—specific goals are better than vague ones, (3) identify a series of learning experiences that will help students progress from what they know to the desired learning goals, (4) devise strategies to monitor learning progress—i.e., formative assessment, and (5) provide feedback about learning progress.

Activity Overview

This Activity that Works helps prospective teachers learn how to use the Interactive (or Question Raising) Approach (see Chapter 6), drawing on the science topic of floating and sinking. The first phase in the teaching model is for the prospective teachers to clarify their own ideas about floating and sinking. In the first activity, an exploration, students explore the floating properties of a variety of objects. They are then invited to raise questions (related to floating and sinking) to which they would like answers. The instructor collects the list of questions, and with the methods students selects some for investigation. The instructor then asks the methods students to propose possible answers to the selected questions, and records them. The methods students, with help from their instructor, devise ways of obtaining information that may help them decide on answers to the questions. They then collect the information, which may require a number of investigations and attempts. After the methods students have collected the information, it is pooled. The students, with instructor assistance, weigh the proposed answers against the evidence. This may lead to students (or the instructor) proposing new questions to investigate and evaluate. The process continues until everyone agrees that a satisfactory answer has been obtained and the methods students understand the answer. The instructor then gives the students a problem to work on that involves application of the science ideas learned.

The learning goals for this floating and sinking Activity that Works include helping methods students build understanding of science teaching, science concepts, and the nature of science:

Pedagogical Knowledge Objective. Build a range of inquiry strategies that support and structure elementary students' learning across a unit of work in science using the Interactive Approach.

Science Content Knowledge Objective. Understand that floating properties are related to the relative densities of the object and fluid, and can be explained as a pair of balanced forces,[1] which can be calculated by the formula: Upthrust = Weight of displaced water (note: this is a concept typically addressed in the upper elementary grades).

Nature of Science Objective. The quality of scientific explanations is judged on the extent to which they are perceived to fit with the available evidence.

To implement the floating and sinking investigation in your methods course, you will need to consider the following:

- *Timeline.* Allow 30 to 60 minutes per session, over four to six sessions.
- *Materials.* Most materials are everyday floating and non-floating items such as pieces of wood, plastic, metal, stone, paper, aluminum foil, and cork of varying sizes. The items must include an apple and a

---

1. A necessary prerequisite for this concept is an understanding of balanced and unbalanced forces.

washed potato. Basins for holding water will be necessary, and several smaller containers would be advisable. Other items needed are a plastic cutting board, vegetable peeler, apple corer, knife (take care with the type of blade), small cookie cutter, cooking salt, measuring cylinder (20 to 100 ml), spring scales, and electronic kitchen scales (measuring grams or ounces). If running water and sinks are not available, use any suitable water storage containers—preferably with a tap outlet, and buckets to catch drips.

Activity

If you are unfamiliar with the Interactive Approach, review it before proceeding (see Chapter 6).

We introduce the floating and sinking Activity by explaining that we will be doing a field-tested example of an inquiry investigation using a teaching model, the Interactive Approach, to structure the learning. The investigation will be spread over several weeks, and with some modifications can be adapted for use in lower, middle, or upper grades. We tell the prospective teachers that although they will be working as students, they should reflect on both what and how they are learning as students; and from a teacher's perspective, on what the instructor is doing, and why. Below is a brief outline of the main steps in the Activity that Works, aligned to the steps of the Interactive Approach. Steps one to three must be completed in the one session.

Step One: Preparation

It is important for the prospective teachers first to clarify their own ideas about floating and sinking, in particular, what is understood by the term "floating." We provide several instances (photographs or sketches are helpful), and ask the prospective teachers whether they consider each to be an instance of floating (or sinking), and why (see Osborne & Freyberg, 1985 for a discussion of Interviews and Instances and Events). Suitable instances (adapted from Biddulph & Osborne, 1984a, 1984b) we use are a person swimming, a stone on the seabed, a piece of wood on the surface of the water, a boat, and a submerged submarine. Another possible instance is an insect "walking" on the surface of water; however we only use this instance with a very capable group, because it introduces a different concept (water tension) and is therefore neither an instance of floating nor sinking. We ask the prospective teachers to say why they think each is an instance of floating or sinking, but do not press for an explanation of floating and sinking. This is deliberate, because if we asked for an explanation at this point, they would attempt to recall the correct scientific explanation, and many of them would be fearful of failing. After discussing the different instances, we help the methods students arrive at a common idea of when an object is floating.

We ensure that this step does not become too lengthy, as it has the potential to become boring. A time of 15–20 minutes is optimal.

Step Two: Exploration

We organize the prospective teachers into groups of three, and assign individual roles of equipment manager, reporter, and group director. If this method of small group organization has not been modeled before, we explain it explicitly. If it has been used on other occasions, we remind the prospective teachers of the value of assigning group roles in cooperative learning situations (Australian Academy of Science, 1994; Biological Sciences Curriculum Study, 1996). We briefly outline the task: first to predict whether each object will float or sink in water, and then to test the predictions. It is the group director's task to ensure that each object is tested in turn, with everyone having a hand. The reporter keeps a record of each prediction and the test outcome. The equipment manager collects, from the equipment depot, a basin of water and a set of objects (this should include but not be limited to a piece of aluminum foil, a stone, a paper clip, a cork, a plastic knife, a screw or nail, an apple, and a washed potato).

While the prospective teachers are doing the activity, we move from group to group checking what they are doing, ensuring records are being kept, and issuing challenges like, "Can you get your piece of aluminum to sink?"

This is only a brief activity, so we do not let it go on too long (10–15 minutes is ample). At the completion of the activity, the equipment managers return the team's objects to the depot. At this point, we take a quick consensus of the results. The reporters provide their group's findings. Although there is a great deal of consistency in the findings, the aluminum, paper clip, apple, and potato usually provide some variation in predictions and results. We are careful to ensure that this feedback component is also fairly brief (depending on the number of groups, about 10 minutes).

The pedagogical principle here is to ensure that all students have a common experience that is readily related to past experiences. At this stage, we are not interested in scientific explanations. The activity should also provide a springboard for the next step. These points need to be made explicit to the prospective teachers. A helpful routine is to use a physical signal to indicate that the methods students should now shift from "thinking like a student" to "thinking like a teacher." The signal can be asking the methods students to change metaphorical hats by removing their "student hat" and putting on their "teacher hat"; or the instructor taking a deliberate sideways step while saying, "An aside. Now we will look at what we have done from a teacher's view." We draw their attention to what the instructor did during the activity, and the type of interaction that occurred. For instance, consistent with our reflection orientation (see Chapter 3), we ask questions like, "What was I doing then as teacher?"

"What interaction pattern occurred?" "Why did I do it that way?" (We find that methods students tend not to be aware of teacher actions, let alone reasons for the actions, so we often have to tell them.) This component may include the methods students keeping journal notes, or constructing a class summary of teacher and student actions on the whiteboard. Once finished, a physical signal like changing hats or reversing the sidestep is again used to return to the activity and "student mode."

Step Three: Students' Questions

We now invite the prospective teachers to raise questions that emerged from the investigation, to which they would like an answer. We find that this can be a productive time to introduce the "think-pair-share" routine (see Chapter 6), as a safe way to encourage all students to participate without feeling publicly embarrassed if they suggest something showing a lack of knowledge (or as children would put it, saying something stupid). As each question is raised, we record it on the whiteboard or similar, without comment. Exceptions would be questions that are obviously rhetorical (e.g., "Was that really a stone?"), or so simple that they are best addressed immediately (e.g., Was the water fresh or sea water?). That is, we want to record questions that might lead to subsequent investigations. Some questions may not be worded well, but could potentially lead to an investigation. In such cases, the methods student asking the question could be helped rephrase it. For example, Faire and Cosgrove (1990, p. 20) report an interaction between a teacher and a child after an exploration of various bones:

ANNE: Are bones different shapes?
TEACHER: Can we answer that by looking at the bones on display?
ANNE: Yes.
TEACHER: Is that the question you really wanted to ask, or was it something else?
ANNE: It was about the shape of our bones.
TEACHER: Oh—which bones?
ANNE: The ones in our back.
TEACHER: O.K.
ANNE: Why are the bones in our back a triangle shape?

It is not unusual for 15 to 20 questions to be raised. Typical questions include:

- Why is some of the wood under the water when it is floating?
- Why did the paper clip sometimes float, but then suddenly sink quickly?
- Would the cork still float if we had a very large piece of cork?
- Why did the apple float and the potato sink?

- Why did the aluminum foil sink when we pushed it under, but stayed floating when crumpled in a ball?

We then discuss with the prospective students which of the questions would be the most productive for further investigations and understanding floating and sinking. Note that the majority of the above questions are "why" questions—that is, they ask for an explanation (elementary students sometimes use the expression, "How come...?"). Harlan (1985) points out that "why" questions should be investigated once elementary students have gained some facility with the teaching approach, so when using the approach initially it is better to start with questions that focus on variables that may be changed during an investigation. However, since the teaching approach is particularly powerful in helping elementary students seek explanatory ideas that relate to important science concepts, we prefer our methods students to experience this aspect in our Activity that Works.

The process we follow is to ask our methods students which question/s are the most interesting and puzzling to them. We have found repeatedly that "Why does the apple float and the potato sink?" is one of those nominated. If it is not, we include it with a comment like, "I think this is a real puzzler." We typically allow only two or three questions to be nominated. We then look at each short-listed question with the view to selecting one for further investigation. Since this is the question that we want to be selected for further investigation,[2] careful negotiation is sometimes required. However, we also remain sufficiently flexible to let go of this question if our students do not warm to it, and select another that will focus on similar concepts. Sometimes a clinching comment can be, "Well, I think this question is the best one, because when we get an answer to it, many of the other questions will also be answered."

The pedagogical principle that we explicitly mention here is how student-generated questions provide windows into what the students already know, and what they do not know. We also mention that more than one question can be selected for investigation, but this increases management issues. We only ever choose one question in the methods course, as this simplifies the process of highlighting the pedagogy.

Step Four: Students' Possible Answers and Planning Investigations

We usually do this step in the next class session. Assuming that the chosen question is "Why does the apple float and the potato sink?" we now invite

---

2. This question is less threatening to those methods students who are fearful of physics (in this case, density and Archimedes' Principle), but also leads to successful investigations into the science concepts those students fear.

the prospective teachers to offer what they think may be possible answers to the question. We find the think-pair-share routine very useful here: many methods students are hesitant to put an idea forward because they are used to normal school transactions where the "right" answer is expected. Note that the phrasing of the questions is important in trying to reduce the expectation of providing a scientifically correct answer: "What do you think an answer to the questions might be?" Two keys components of the question are "you think" and "might."

Answers suggested depend on the sophistication of the prospective teachers, so it may be necessary for the instructor to introduce a few answers if they are not forthcoming. For example, "Some children have told me they think the apple floats because there is air around the seeds"; or "Some children have told me that they think the apple floats because the skin is watertight, and no water can get in to make it sink."

We accept all tentative explanations, and make a list of them. Sometimes it may be necessary to ask a question to clarify what is meant. Often someone in the class will invoke Archimedes' Principle, with simple reference to the term, the formula, or both. We add that idea to the list. Once the list is completed, an explanation that seems plausible to most students is selected.

We then pose the question, "How can we find out whether that is a good explanation?" (Note that we do not ask about finding the "right" explanation.) We encourage the methods students to think of an investigative test that could be done, rather than merely looking the answer up. The think-pair-share routine can again be used here within the groups of three, as students will work in those groups while conducting the test(s).

Suppose the selected explanation was "the apple floats because there is air around the seeds." A suitable test would be to core the apple, and then see if it sinks. This is a simple test, so could be done without further detailed planning, but a more complex test would require planning of detail cooperatively.

The pedagogical principle here is that tentative explanations proposed by the students expose aspects of what they know, and what they do not know, about the topic. Asking them to be involved in planning investigations also reveals their grasp of important principles such as conducting a fair test (controlling variables). We explicitly relate conducting tests to the nature of science objective as well: valid evidence (such as through controlling variables) that will either support or contradict tentative explanations needs to be obtained.

Step Five: Specific Investigations

This step may be spread over several class sessions, depending on the tests conducted.

Once the plan of the test investigation is determined, the prospective teachers, in their groups, carry out the planned test. Once the tests are completed, the groups report their results to the class immediately, and the explanation being investigated is evaluated in terms of the test results. In this example, the cored apple still floats, so the possible explanation about air around the seeds can be eliminated.

The next most plausible answer is then selected, tests devised, carried out, and evaluated. For instance, the answer "the apple floats because the skin is watertight" is tested by peeling the apple, and seeing if it sinks. This explanation is then eliminated. We always tell the prospective teachers that if a new explanation occurs to them, they should feel free to raise it for consideration.

Eventually (or even initially), an explanation about density or Archimedes' Principle will be selected. Both are acceptable explanations, but require more detailed planning of the tests. We prefer testing both explanations so students can see that there are different ways of explaining the same event, depending on the theoretical stance taken (an important aspect in the nature of science). We find that the amount of science experience the prospective teachers have determines the extent that they are able to relate to these ideas and plan tests. However, more science experience is no guarantee that they will understand the ideas being explored. We usually find it necessary to make several suggestions during the planning of the tests to ensure that the tests are testing what is intended, and that useful data will result.

Students with only an intuitive grasp of density may use an expression like that offered by a grade 6 child: "the stuff in the potato is packed in more tightly than the stuff in the apple." This explanation can be tested by cutting identical sized pieces (i.e., pieces with the same volume) of potato and apple (using the cookie cutter) and weighing them. The piece of potato weighs more. The weights can be compared to the weight of the same volume of water. Obtaining the same volume of water using displacement is a good way of introducing this idea if the students are not already aware of it.

Prospective teachers are often drawn to Archimedes' Principle first because they have heard of it; it has the status of being named after someone, and it sounds complicated—so by implication it must be right. We find it is usually best to gently move them away from investigating this initially, so they can build confidence with investigating their own ideas that make intuitive sense. Students will not readily understand Archimedes' Principle unless they understand that opposing balanced forces result in no movement, and opposing unbalanced forces result in movement. We find it is usually necessary to revisit this idea using concrete examples before proceeding. Just because a student proposed an idea and the others agreed with it does not mean that everyone understood it, and could see how it related to the question being investigated. That is, it is important for everyone to actually understand what the explanation suggests: many will say that they understand it when in fact they do not. What they are really saying is that they think the explanation sounds plausible because it vaguely links to some remembered science, and since it was suggested by a colleague deemed as academically capable it must be right. You may wish to revisit the view of learning outlined in Chapter 1 (particularly the degree of fit section) to understand what is happening here. Further, even the student(s) who suggest Archimedes' Principle as an explanation may not fully understand it. For instance, we find that few students can say why there is an upthrust force, as Archimedes' Principle does not state this. Yet without understanding the origin of the upthrust force (water pressure due to the weight of water), Archimedes' Principle is insufficient of itself in answering the question.

It is best to test Archimedes' Principle using the traditional method of measuring the weight of an object using spring scales before immersing it in water, and then while it is immersed using displacement to collect the volume of water for determining the weight of water displaced by the object. While a displacement jug makes for greater accuracy when doing this, a tin can or glass can be quite adequate. We find that this is also a good opportunity to provide instruction on accuracy of measurement—the measurements taken in these tests will result in volumes and weights that are not precise. Some students intuitively expect "Upthrust = Weight of water displaced" to have identical numbers on each side, and do not consider errors in measurement.

Once an explanation has been found to be supported by the tests, the prospective teachers should be invited to check other sources for further information (such as books, Internet), and possible confirmation. During this process and the subsequent reporting phase, we find it necessary to work with individual students to help them make sense of the tests conducted, the test outcomes, how the test outcomes relate to the question, and the information they are finding from other sources: students do not necessarily put it all together intuitively. We find it is best to conclude this step by asking each student to write in their own words the best explanation to the selected question, including an outline of the explanation and how it actually answers the question. They should also include reference to other relevant information sources. Sometimes we ask them to comment about how their process of determining the best explanation might be like or different from how scientists work.

There are several important pedagogical principles here. Research has shown (e.g., Tasker & Freyberg, 1985) that students often do not understand investigative procedures unless they are involved in working them out. So it is important for them to be involved in planning investigations, and understand why things are being done as they are. Just because one person in the group, or the instructor, explains how something should be done does not mean that all understand what to do or why. Further, students are accustomed to finding answers in books and the like, and do not necessarily relate this information to the task at hand, hence the expectation that they actively evaluate test results and work toward an explanation. Finally, the sequence of proposing and conducting investigations to test ideas, and accepting or rejecting ideas on the basis of the available evidence provides a worked pedagogical example of aspects of the nature of science (see the nature of science objective outlined at the start of this Activity).

Step Six: Problem Solving

From the previous step, prospective teachers should have demonstrated an understanding of density and/or displacement in their explanations. However, they may not fully grasp the concept if not asked to solve a problem that requires the application of the newly learned information for its solution. Two problems we have used are:

- make the potato float;
- make a clay boat that will carry a cargo equal to its own weight.

We usually pose only one of these problems, and ask the prospective students to work in their assigned groups of three to solve it. We ask them first to plan how they might solve the problem, without requiring a lot of detail. They may need a hint if stuck (for example, if the explanation focused on the relative densities of the potato and water, we might ask a series of questions, "What if you could increase the density of the water?... What would happen if the density of the water were greater than the density of the potato?... What could you do to increase the density of the water?"). They should be able to reach a solution within one class period. We conclude this step by asking our methods students to explain why their solution worked. For instance, making the potato float by dissolving salt in the water can be explained in terms of increasing the density of the water, or in terms of increasing the weight of the water displaced.

The pedagogical principle here is that students need to work with information several times in order to understand it (Nuthall, 1999, 2001). This step is also a small inquiry investigation within a larger one.

Step Seven: Reflection

We conclude this Activity that Works by asking the prospective teachers to review the initial set of questions asked, and select those that have been answered by the explanation generated for the selected question. They should find that several of the questions have been addressed. We ask them to explain the answer to some of these questions: depending on class schedules this can be done as a whole class exercise (individuals explaining to the class), as a small group task (each group constructs an explanatory poster for display), or as an individual home task in their journals. Finally, we invite them to reflect on what they have learned during the investigation: in terms of science, pedagogy, and the nature of science. As outlined in Chapter 6 the pedagogy, generically called inquiry, can be looked at in terms of routines, techniques, strategies, and teaching models, not to mention the interpersonal interactions between instructor and student. It is useful to highlight the importance of the teacher–student interactions that make this teaching approach so effective and that give it its name.

The pedagogical principle in this step is that elementary students should engage in reflection about their own learning, and the progress they have made toward the desired goals. The information from these reflections can also be a useful data source for the assessment of learning.

Making the Activity Work for You

In our experience, this Activity that Works is one of the most important experiences we provide for prospective teachers in our methods class. Many come with the idea that planning a unit of instruction needs to done completely in advance. In the Interactive Teaching model, while there is some certainty about what will happen, there is also uncertainty: will the desired question/s be raised; will the investigations planned be similar to those anticipated; can resources for the investigations be organized in time? Seeing their instructor manage this uncertainty provides an effective model for the prospective teachers, and engaging in this authentic inquiry experience helps them see the learning benefits for students. Further, the investigation is one that they could easily adapt for their own teaching in elementary schools. It would work with little alteration in upper elementary grades. Middle and lower grades would require greater changes, as elementary students from these grades would not be able to readily grasp explanations like Archimedes' Principle. For middle grades, the density explanation is the most productive to investigate; and for lower grades, investigations that focus on the relative size and weight of objects are most fruitful.

We have found for some methods courses that the whole investigation cannot be carried out because of time constraints. In such instances, we have found that some of it can be compressed into two periods—mainly by shortening the sequence of investigations to test the various ideas. The students devise the tests, but they only conduct some of them. For the remainder of the tests, we tell them results that "others" have obtained. Provided adequate explanations are provided, this can be a reasonable compromise.

Tools for Implementing the Activity

We have used excerpts from Biddulph and Osborne (1984a) to help our prospective teachers understand that the key to the Interactive Approach is not that it just has a series of steps to follow, but that student learning is highly dependent on the type of interaction between teacher and students. We reproduce useful sections below:

Purposes of the Approach

There is nothing unusual about teachers interacting with children. It occurs in classrooms all the time. However, we use the term "interact" in the sense of an interchange of talk among people who respect each other's ideas. From a teacher's point of view, this begins with a genuine desire to know what a child thinks (and why). The main purposes of an interactive approach to teaching are:

i. to identify children's present ideas and questions,
ii. to provide children with stimulating experiences either to confront and explore those ideas or as a basis for developing ideas; in either case the experiences should help children raise questions,
iii. to help children develop, clarify, modify, and extend their ideas through seeking answers to questions they are interested in (or can be interested in) or through checking proposed answers,
iv. to encourage children to reflect critically on how they came by an idea and whether it is a sensible and useful one,
v. to assist children develop the skills they need to ask better questions, plan and carry out investigations, and construct and communicate ideas,
vi. to help children realize that explanations of why things behave the way they do are frequently not "right" or "wrong" but are rather consistent with the evidence or inconsistent, useful or less useful, plausible or not plausible, intelligible or not intelligible, and
vii. to convey to children an awareness that their genuine ideas are valued.

(Working Paper No. 122, p. 10)

Roles for the Teacher

There are several effective roles that teachers can adopt in the course of interactive teaching. These include:

Facilitator of Learning

In this role, the teacher tries to bring children and relevant resources together. The teacher, for example, may suggest seeking the views of an expert who is invited into the classroom, or a child may obtain information from a parent or neighbor whose work is relevant. A teacher directing a child to a particular book or equipment would be other examples.

Resource Person

At times, the teacher may have information that a child is seeking. In that case the teacher can act as a resource person in the same way as a knowledgeable parent, expert, or other member of the community, rather than turning a question back on a child and sending him or her off in another direction to look for an answer.

Naïve Fellow Investigator

In this role, the teacher expresses ignorance of an explanation or situation. In a sense the more genuine the ignorance and the more willing the teacher is to learn from the student the better! For example:

CHILD: I think that big block of ice is colder than the little block.
TEACHER: Colder, mm, I can't say I've ever checked it myself. I wonder if you are right. Is there any way we can find out?

CHALLENGER OF IDEAS

Here the teacher deliberately but sensitively challenges those ideas expressed by the child that are inconsistent with evidence, not useful, not clear, and so on. For example:

CHILD: I think spiders chew up flies for their food.
TEACHER: Chew them up? Uh-huh. Someone in that other group said the spider sort of sucks the "blood" out of the fly.
CHILD: Ohh yeah, could be. I'm not sure if they have teeth. No, they don't have teeth; they have those nipper things.

The challenge posed by the teacher in this role has the effect of revealing the child's commitment to his/her present view and perhaps helping the child to clarify or reconsider the view held. A further example is:

SCOTT: [Pointing to cross-section of slice of tree trunk] I wonder why trees do have those rings in the wood?
MICHAEL: I think it might be 'cause of the bark.
TEACHER: Bark, mm; how do you mean, the bark?
MICHAEL: Well, as the tree gets fatter the bark goes further out, and where it was last time it leaves a mark 'cause bark is a bit softer. You can feel it on a twig.

TEACHER: Yes, I see what you mean. I think we will have to look at that idea.

A final example illustrates the diverse nature of the role of the challenger of ideas.

TEACHER: John [one of the children] was wondering why trees have leaves.
CHILD: Um ... Food ... I think.
TEACHER: I thought for a moment you were going to say that the leaves make food for the tree.
CHILD: Yeah, they do, sort of.
TEACHER: Mm, how do you mean?
CHILD: Well, they drop down onto the ground and kind of go rotten, make manure and that for the tree to feed on.

"Leaves make food for the tree" is a common expression used by many teachers: the leaves use light, and the energy-poor materials water and carbon dioxide, to produce energy-rich materials that are used, in part, as "food" for the tree. Unfortunately, the above child has interpreted the words in quite a different way. As a challenger of children's ideas, a teacher should not always take children's responses as face value but should sensitively explore what they have in mind (Working Paper No. 122, pp. 36–38).

Working Paper No. 129 contains a lot more detail about the respective steps in the Interactive Approach, but is too voluminous to reproduce here.

More to Explore

The best sources to explore the Interactive Approach are no longer in print. However, they are held in many university libraries in Australia and New Zealand (e.g., Central Queensland University, Rockhampton, Australia, and Waikato University, Hamilton, New Zealand), and a few in the United States and in the United Kingdom. You can obtain copies from these libraries if you have sufficient lead-time:

Biddulph, F., & Osborne, R. (Eds.). (1984). *Making sense of our world: An interactive teaching approach*. Hamilton, New Zealand: Science Education Research Unit, University of Waikato.

Faire, J., & Cosgrove, M. (1990). *Teaching primary science*. Hamilton, New Zealand: Waikato Education Centre.

Harlen, W. (1985). *Teaching and learning primary science*. London: Harper Row.

Other more readily accessible references to explore include:

Chin, C., & Osborne, J. (2008). Students' questions: A potential resource for teaching and learning science. *Studies in Science Education, 44*, 1–39.

Klentschy, M. (2005). Science notebook ESSENTIALS. *Science and Children, 43*(3), 24–27.

Turner, J. (2006). Thinking about students' questions. *Science Scope, 30*(3), 51–54.

Weizman, A., Shwartz, Y., & Fortus, D. (2008). The driving question board: A visual organizer for project-based science. *Science Teacher, 75*(8), 33–37.

References

Australian Academy of Science. (1994). *Primary investigations*. Canberra, Australia: Australian Academy of Science.

Biddulph, F., & Osborne, R. (1984a). Floating and sinking. In F. Biddulph & R. Osborne (Eds.), *Making sense of our world: An interactive teaching approach* (Working Paper No. 130). Hamilton, New Zealand: Science Education Research Unit, University of Waikato.

Biddulph, F., & Osborne, R. (1984b). Pupils' ideas about floating and sinking. *Research in Science Education, 14*, 114–124.

Biological Sciences Curriculum Study. (1996). *Science for life and living*. Dubuque, IA: Kendall/Hunt.

Faire, J., & Cosgrove, M. (1990). *Teaching primary science*. Hamilton, New Zealand: Waikato Education Centre.

Harlen, W. (1985). *Teaching and learning primary science*. London: Harper Row.

Nuthall, G. (1999). The way students learn: Acquiring knowledge from an integrated science and social studies unit. *Elementary School Journal, 99*, 303–341.

Nuthall, G. (2001). Understanding how classroom experience shapes students' minds. *Unterrichts Wissenschaft, 29*, 224–267.

Osborne, R., & Freyberg, P. (1985). *Learning in science: The implications of children's science*. Auckland, New Zealand: Heinemann.

Tasker, R., & Freyberg, P. (1985). Facing the mismatches in the classroom. In R. Osborne & P. Freyberg (Eds.), *Learning in science: The implications of children's science* (pp. 66–80). Auckland, New Zealand: Heinemann.

Activities that Work 3

Inquiring into Guided and Open Inquiry
Insect Study

Introduction

According to many science education policy documents, elementary students need a chance to plan and carry out investigations. We believe that, to support elementary students in such inquiry activities, future teachers need their own chances to plan and carry out a science investigation. In Chapter 6 we discussed instructional strategies, such as the investigation strategy, that can support inquiry-based science instruction. In this Activity that Works, we demonstrate how such strategies might play out in the methods course. In this Activity, we support students to launch an investigation with living organisms, starting with guided whole class inquiry at the beginning of the semester, and moving the students to a full and open independent inquiry (National Research Council (NRC), 2000) by the end. Our aim is to help the prospective teachers build their PCK for instructional strategies while they practice scientific inquiry and learn about insect life cycles.

We like to begin the elementary science methods course with a life science experience, knowing that preservice teachers tend to feel most comfortable and confident with biological concepts. However, we take them out of their comfort zone a bit by using an insect, the darkling beetle, as the study organism. Although the initial reaction of our students to the mealworm—the larval stage of the darkling beetle—is disgust, we believe that gaining familiarity with the insects takes away those icks and yuks (Ballou, 1986) and helps the future teachers prepare to encounter life science in their classrooms. Mealworms are inexpensive, easy to acquire at many pet stores and bait shops, and require no specialized equipment to raise and maintain. A group of organisms can be provided to methods students in plastic tubs that each can take home for investigating. Thus the darkling beetle is an ideal organism for independent, student-directed inquiry in the methods course.

Activity Overview

The mealworms insect study Activity that Works takes place in three phases. In Phase I, the instructor gives methods students a container of mealworms and guides an inquiry of insect life cycles. In Phase II, students design and carry out their own investigations with mealworms, keeping track of their work in their science notebooks (Campbell & Fulton, 2003). The individual inquiries into mealworms take place outside of the methods course as homework, however some class time is set aside for students to share the progress of their investigations and compare their notebooks. In Phase III, students present the results of their investigations through a poster at the class Mealworm Science Conference (see Olson & Cox-Peterson, 2001). They also write a final notebook entry to reflect on their experience.

The learning goals for this insect study Activity that Works include helping methods students build understanding of science teaching, science concepts, and the nature of science:

Pedagogical Knowledge Objectives. To understand that the degree of student-directedness in inquiry can vary from teacher-guided to open inquiry. Also to understand that teachers and students should follow ethical guidelines for using living organisms in the science classroom.

Science Content Knowledge Objectives. Plants and animals have life cycles that include being born, developing into adults, reproducing, and eventually dying. The details of this life cycle are different for different organisms (NRC, 1996, p. 129, Grades K-4 standards).

Nature of Science Objectives. There is no one scientific method. Instead, "scientists use different kinds of investigations depending on the questions they are trying to answer. Types of investigations include describing objects, events, and organisms; classifying them; and doing a fair test (experimenting)" (NRC, 1996, p. 123).

To implement the mealworm inquiry in your methods course, you will need to consider the following:

- *Timeline.* Provide 20 minutes of class time for each of the first three class periods, two "research meetings" of about 20 minutes each, about one-third and two-thirds of the way through the semester, and an entire class period for the final science conference.
- *Materials.* Each methods student will need a science notebook, a hand lens, and a small plastic tub of darkling beetles (about 5–10 mealworms and other life stages). Other materials supplied by individual student investigators as needed.

Activity

I often begin Phase I of this Activity that Works on the first day of class. After discussing the goals and expectations for the course, I ask methods students to answer some questions about the nature of science, modified from the Views of the Nature of Science (VNOS) instrument (Adb-El-Khalick, Bell, & Lederman, 1998). In particular, I ask them about the role of experiments in science, whether the development of scientific knowledge requires experiments, and what they think constitutes an experiment. We revisit their answers to these questions during the insect inquiry.

During this first class period, I also want the preservice teachers to experience a natural phenomenon, and darkling beetles always create a sensation. While the methods students are seated at their tables, I place a covered container of organisms at each table. I am careful to select a group of beetles that contains the three visible life stages of the insect—larva (mealworm), pupa, and adult (see note at end of this chapter). Because I grow the mealworms in our science methods lab year-round, I always have plenty from which to choose. When I have everyone's attention, I ask them to choose someone at their table to uncover the container. Their reactions are predictable—"icks" and "yuks" predominate. Sometimes I read aloud Ballou's story (1986), which encourages the prospective teachers to turn their "icks and yuks" into "oohs and aahs" and become positive role models for their future students.

Once the initial reaction to the insects has subsided, I ask the preservice teachers to spend a few minutes observing the organisms in their containers. We start a class chart of their observations. The students recognize that there are different things in their containers (they usually call them "worms" and "bugs"), but seldom does anyone suspect that these are different life stages of the same organism. That's when I introduce the question that will guide our first inquiry into mealworms: "We are going to inquire about these organisms with this question—What changes will the organisms exhibit as they grow? What can we do to find the answer to this question?" I ask. The students respond that we can observe the organisms and take notes about what happens over time. At this point I distribute a container of insects to each student:

> These organisms are now under your care. Please read the homework assignment about how to be responsible for living organisms in the science classroom (National Science Teachers Association, 2005). Watch your organisms over the next several days and take some notes about what you observe. We will start our next class with your observations.

Over the next couple of class periods, students share their observations. Some witness the larval mealworms shedding their skin, others discover pupa and adults. However, most do not have scientific words for the different life stages, nor do they realize that every creature in the container is a darkling beetle. We add all of these observations to our class chart, using the students' own words. A few students inevitably surmise that the different creatures they see are actually life stages of the same organism. When they are ready, I tell them that the organism is a darkling beetle and label the four life stages involved in its complete metamorphosis. To summarize this phase of their inquiry, I ask students to draw a life cycle diagram in their notebooks, and then compare the darkling beetle life cycle with life cycles of another animal and of a flowering plant.

At this point, students are ready to discuss their earlier responses to questions about the experimental nature of science. We compare answers and agree that experiments involve variables and fair testing. Then we discuss whether or not our insect inquiry constituted an experiment (i.e., did they manipulate a particular variable to observe its effect?). After some discussion, students are usually ready to accept that, although they were engaged in a scientific inquiry into insects, they did not conduct an experiment. They agree that sometimes scientists observe without experimenting to come to conclusions. I point out that this is an important idea about the nature of science, which they reflect upon in their science teaching notebook.

With this brief introduction to darkling beetles and the nature of science through a teacher-guided inquiry, we enter Phase II of this Activity that Works. During this phase, each student conducts an open inquiry about the insects. I tell them that instead of having my guidance, they will take ownership for the inquiry. Moreover, different from our first observational inquiry, they will be conducting an experimental inquiry, requiring a fair test. As a class, we brainstorm questions that could guide their inquiries. I pass out and review the Inquiry Project Assignment (see "Tools" section of this chapter).

I send the methods students home to read articles about questions and fair tests (e.g., Capobianco & Thiel, 2006; Cothron, Giese, & Rezba, 2006), and to think about other questions they may want to investigate. Over the next week or two, they complete the Investigation Planning Sheet (see "Tools"), derived from Harlen (2001). They email me their plans, which I review. When I have OK'd their investigation question and plan, they are ready to proceed. During class, we discuss the Harlen chapter on planning an investigation, noting similarities and differences with their investigations. I also share examples of plans made by actual elementary students, which we examine in terms of the children's fair testing abilities.

Methods student investigations have included studying insect behavior (e.g., responses to noise, light, and temperature) and eating habits (e.g., preferences among various fruits and vegetables) as well as comparing such results for different life stages. They collect their data outside of class over a number of weeks. Later in the semester, we take about 20 minutes during two different class periods for students to participate in a research meeting. During this time, they share their progress on their investigations. They compare their observations and raise questions about their own and each other's investigations. Also during class we discuss a continuum of inquiry that can occur in elementary science (NRC, 2000), how scientists use different methods to answer different types of questions, and how to support elementary students during open inquiry. We also work together to generate a scoring rubric for their final projects that will take into account the essential features of classroom inquiry about which they have been reading (NRC, 2000). A sample scoring rubric is included in the "Tools" section of this chapter.

Phase III of this Activity that Works includes the Mealworm Science Conference that takes place on the final day of the course, and self-evaluations. Earlier in the semester, I tell students that scientists present their work to each other in conference settings as part of a community to scrutinize and understand each other's work. After hearing that they are required to make a poster of their inquiry to share at our class Mealworm Science Conference, students develop poster guidelines and summarize their work accordingly. At the Conference, half of the students stay at their posters while half walk around and interact with the researchers about their studies. Then they switch so each student has the chance to play each role. The Conference is an event to which I often invite other elementary education majors, faculty colleagues, administrators, and local teachers to celebrate the investigations that have taken place.

To reach closure on the semester-long insect inquiry, I ask students to complete a science notebook entry (see "Tools") in which they reflect on what they learned about insects, about science teaching, and about the nature of science. They also use the class-generated scoring rubric (see "Tools") to score their performance. Although I also grade their projects, I have found that, in nearly all cases, my grades align closely with how they score themselves. The insect inquiry project and the associated Mealworm Science Conference provide a sense of accomplishment to the prospective teachers, who feel ready to support their own students in guided and open inquiry projects.

Making the Activity Work for You

The study of the darkling beetle is an interesting journey in the methods course. Students often start out with negative feelings toward the insects and negative perceptions of science and their own abilities to succeed in science. The journey from guided to open inquiry changes their minds. Students take ownership of their organisms and their inquiries. They become confident in their abilities to pose researchable questions, plan investigations, and collect and analyze data. They see the value of engaging elementary students at various points on the inquiry continuum.

The insect inquiry project is relatively easy to undertake, does not require much in the way of specialized equipment, and illustrates important ideas in biology (life cycles) and nature of science (ways of doing science). Student reactions to their own science learning typically are enthusiastic. However, some are skeptical about their abilities to succeed using a similar activity with elementary students. We take some time to read examples from teachers in the field (e.g., Tower, 2000) that demonstrate how teachers work through barriers they encounter to inquiry-based instruction in science and in other subjects, and discuss how to find the supports needed for success. We offer the insect study as an Activity that Works in the elementary science methods course in terms of preservice teachers developing their knowledge and self-efficacy as science learners and teachers. We support your use of the Activity with the Tools for your own methods course.

Tools for Implementing the Activity

Notes on the Darkling Beetle Life Cycle

Darkling beetles follow a life cycle known as complete metamorphosis, consisting of four distinct life stages. A female adult beetle lays eggs, which are as small as the period at the end of a sentence. After a couple of weeks, tiny larvae emerge from the eggs. The larvae are known as mealworms (although they are not worms at all) and have the characteristic six legs of an insect. They are quite active and interesting to watch. The larvae molt (shed their skin) repeatedly as they grow. When they molt, they initially appear a translucent white, but gradually darken to brown again. This may occur a number of times over a 3-month or so period, until the larvae reach about 2 cm. The final larval molt reveals the pupa, which barely moves unless disturbed. In 2 or 3 weeks, the pupa splits open and an adult beetle emerges. It is white at first, but after a day turns a dark brown. The adult beetles mate, the females lay eggs, and the cycle repeats itself. (See University of Arizona Center for Insect Science Education Outreach, 1997.)

Individual Inquiry Project Assignment Description

Rationale

> From the earliest grades, students should experience science in a form that engages them in the active construction of ideas and explanations and enhances their opportunities to develop the abilities of doing science. Teaching science as inquiry provides teachers with the opportunity to develop student abilities and to enrich student understanding of science.
>
> (NRC, 1996, p. 121)

An important part of preparing to teach science as inquiry involves experiencing inquiry yourself. According to the U.S. *National Science Education Standards* (NRC, 1996), inquiry takes place on a continuum of guided to open (also see Colburn, 2000). In class you will participate in an inquiry guided by your instructor. Then you will conduct your own open scientific inquiry. This process will allow you to identify critical issues and anticipate difficulties your own students might experience with guided and open inquiry.

The inquiry project is designed to help you:

- develop abilities of inquiry (*NSES* Standard A: Science as Inquiry) including asking a question, planning and conducting investigations, and using data to construct explanations;
- use science notebooking to make thinking visible and support inquiry;
- understand that the nature of scientific work includes a range of investigation methods;
- develop comfort being in a "place of confusion" as you construct new understandings;
- overcome personal "icks and yuks" related to investigating living organisms;
- learn to use live animals in your classroom responsibly and within guidelines.

Expectations

You will be given a group of live organisms to study. As you carry out your experimental inquiry, you will take on the role of a student/learner, but will also reflect on the implications for your future teaching.

- Each investigator should follow the National Science Teachers Association Guidelines for Responsible Use of Animals in the Classroom (NSTA, 2005). Investigators should provide for the basic needs of their organisms!
- Each investigator will design and conduct a fair test (experiment) to answer a question of his/her own about the organisms. *Before you begin collecting data, you must get your investigation question and plan approved by your instructor* (see Investigation Planning Sheet).
- Each investigator should keep a detailed and accurate account of his/her investigation and data using a science notebook (see Campbell & Fulton, 2003).
- In class, investigators will participate in periodic small-group "research meetings" to share and compare the progress of their investigations, their observations, inferences, and questions.
- Each investigator will present his/her findings with a poster presented

at the Mealworm Science Conference to be held the last week of class (see Olson & Cox-Peterson, 2001).

A Word of Caution

Remember that if, as a student, you experience difficulty, it is time to shift to the role of teacher to decide how you could help ease the student's (your own) frustration. The readings provide helpful examples you can implement as you embark on this project. You should also draw upon your fellow investigators to help address your questions during research meetings held in class. By tackling some of your own stumbling blocks during the inquiry, you'll be better prepared to help your future students do the same. Though things like planning an investigation or keeping a notebook may be difficult at first, you will show improvement and gain confidence over the course of the project. This project should model what you might expect from your own students in terms of conducting the investigation, keeping your notebook, and giving your presentation. Above all, you should model scientific attitudes and habits of mind of inquiry, such as curiosity, skepticism, and respect for evidence.

Evaluation

The class will work together to construct a scoring rubric outlining the evaluation criteria for this assignment. Each investigator will perform a self-evaluation using the rubric. The instructor will also evaluate your work using the rubric. The two scores will be averaged to determine your final grade for the project.

Investigation Planning Sheet[1]

Mealworms Inquiry

Investigator _____

A. My investigation question: _____

_____

                                          Teacher OK _____

B.

What should I change in the investigation? (independent variables)	
What should I keep the same? (controlled variables)	
What I will observe/count/measure/describe: (the dependent variable)	
How much I will count/measure/describe:	
How I will count/measure/describe:	

C. How will the result be used to answer the investigation question?

_____

_____

_____

D. What I need to learn how to do before I can proceed:

_____

_____

E. What I predict I will find out: _____

_____

_____

---

1. Some of the ideas for this investigation planning sheet are adapted from Harlan, 2001.

F. The steps I will take and the materials needed:

Steps	Materials
1.	
2.	
3.	
4.	
5.	

Teacher OK _____

Sample Inquiry Project Self-Reflection and Scoring Rubric

Final Notebook Entry

In your science notebook, please respond to the following questions as a summative reflection on your investigation.

Part One: Your Inquiry Investigation

- What was your question? What observation(s) led you to this question?
- What inferences/answers to your question did you consider?
- How did you design your investigation to come up with an answer? What evidence did you decide to collect? Why? How did you ensure a fair test?
- What did you find?
- What conclusions did you draw from these data?
- What additional information did you gather related to your question? Where did you seek information? How do your conclusions compare to other information you found?
- How certain are you of your conclusion? Are there other possible explanations for your data? What would make you more certain of your conclusions?
- What would you do differently if you were to do your investigation again? What other data/information would you seek? How?

Part Two: Your Science Notebook

- What is one example of a way in which you used your notebook effectively? Tell how this example illustrates the strategies suggested in Campbell and Fulton (2003).
- What is one example of something you could improve upon in your use of the science notebook? Provide suggestions for improvement by drawing on the strategies discussed in Campbell and Fulton (2003).

Part Three: The Nature of Science

- Would you consider your investigation "scientific?" In what sense?
- How do you think your investigation compares to the ways in which scientists go about their work? Provide examples to support your answer.

Scoring Your Work

Once you have completed your final notebook entry, please use the rubric to self-assess your performance on this assignment. Then, in the box below, indicate the level of performance you feel you achieved. In the comments box, provide a rationale for the grade you feel you deserve. Include this in your notebook when you turn it in to your instructor.

Overall performance: • Exemplary • Satisfactory • Needs improvement	Comments:

Inquiry Project Evaluation Rubric

Criteria	Exemplary	Satisfactory	Needs improvement
Asks a question about objects, organisms, and events in the environment	The question should be student generated, coming from prior knowledge or previous experiences. The question should be open-ended and relevant to the topic and should be answerable through an experiment.	The student asks a testable question, but the question may lack focus and need clarification.	Student does not generate a question that links to his/her own prior knowledge and experience and/or the question is not testable through an experiment.
Plans and conducts a simple investigation	Student investigation is well conceived and thoughtfully planned. Fair testing conditions are met. The data collected are appropriate, given the question asked. The student utilizes process skill(s) systematically to collect relevant data. Data are recorded in an organized manner in the science notebook.	The investigation is logical, but lacks a systematic approach, which makes comparison of data difficult; however, relevant data are collected to answer the question. These data are recorded in the science notebook.	The investigation is not designed in such a way to enable collection of data relevant to answering the question and/or does not take into account variables that affect the data.
Employs simple equipment and tools to gather data and extend the senses	All tools* necessary to conduct accurate observations and measurements were used. These tools were employed appropriately and effectively to enhance the research and/or communication of findings. (*includes science notebook)	The student neglects use of appropriate tools that could support data collection/communication and/or does not use tools effectively.	The student neglects use of necessary tools to aid data collection and/or communicate findings.

Uses data to construct a reasonable explanation	Student provides examples of and uses data to construct and support logical explanation(s). Students consider the strengths and merits of the evidence, as well as alternative explanations for their findings. Findings are interpreted in light of existing scientific knowledge and/or findings of others.	The student forms an explanation based on his/her evidence without considering the strengths of his/her data, how it compares to other reliable sources, and/or other alternative explanations for their findings.	The student's explanation is not linked to and/or supported logically by his/her data.
Communicates investigations and explanations	The student presents findings in a way most applicable to the research, using written, verbal, and/or pictorial communication. Student presentation goes beyond simply reporting results to communicate a reflective analysis of the investigation. The student is able to recognize the strengths and limitations of the investigation and propose ways to improve his/her inquiry methods.	The student presents his/her findings, but may simply report results rather than communicate an explanation of those results or the student may not be able to provide an evaluation of the strengths and limitations of his/her investigation and propose ways to improve it.	The student presents his/her findings, but may simply report results rather than communicate an explanation of those results and the student may not be able to provide an evaluation of the strengths and limitations of the investigation and propose ways to improve it.
Demonstrates understandings about scientific inquiry	The student connects his/her investigation to the work of scientists and the nature of scientific knowledge through specific example(s) and explanations.	The student emphasizes the processes of science only or fails to provide examples to connect the investigation to the work of scientists.	The student fails to make a relevant connection between his/her investigation and the work of scientists.

More to Explore

Brown, S. (2006). What's bugging you? *Science and Children, 43*(7), 45–49.
Colburn, A. (2000). An inquiry primer. *Science Scope, 23*(6), 42–44.
Cox-Petersen, A. M., & Olson, J. K. (2001). Promoting puzzlement and inquiry with pillbugs. *Science Activities, 37*(4), 20–23.
Staller, T. (2005). Meet the mealworms. *Science and Children, 42*(5), 28–31.

References

Abd-El-Khalick, F., Bell, R. L., & Lederman, N. G. (1998). The nature of science and instructional practice: Making the unnatural natural. *Science Education, 82,* 417–436.
Ballou, M. (1986). Taking the icks and the yuks out of science. *Science and Children, 23*(7), 7–8.
Campbell, B., & Fulton, L. (2003). *Science notebooks: Writing about inquiry.* Portsmouth, NH: Heinemann.
Capobianco, B., & Thiel, E. A. (2006). Are you UV safe? *Science and Children, 44*(1), 26–31.
Colburn, A. (2000). An inquiry primer. *Science Scope, 23*(6), 42–44.
Cothron, J. H., Giese, R. N., & Rezba, R. J. (2006). *Students and research: Practical strategies for science classrooms and competitions* (4th ed.). Dubuque, IA: Kendall Hunt.
Harlan, W. (2001). *Primary science: Taking the plunge* (2nd ed.). Portsmouth, NH: Heinemann.
National Research Council (NRC). (1996). *National science education standards.* Washington, DC: National Academy Press.
National Research Council (NRC). (2000). *Inquiry and the national science education standards.* Washington, DC: National Academy Press.
National Science Teachers Association (NSTA). (2005). *Responsible use of live animals and dissection in the science classroom.* NSTA Position Statement. Retrieved July 1, 2009, from http://www.nsta.org/about/positions/animals.aspx.
Olson, J. K., & Cox-Peterson, A. M. (2001). An authentic science conference. *Science and Children, 38*(6), 40–45.
Tower, C. (2000). Questions that matter. *Reading Teacher, 53,* 550–557.
University of Arizona Center for Insect Science Education Outreach. (1997). *Using life insects in elementary classrooms for lessons in life.* Retrieved July 1, 2009, from: http://insected.arizona.edu/rear.htm.

Activities that Work 4

Eliciting Student Ideas
The Human Body

Introduction

The view of learning discussed in Chapter 1 highlighted the importance of an instructor being aware of students' prior knowledge about a science topic to be studied, and then devising a scaffolded instructional sequence that takes the students from what they know to the desired learning goals. In this Activity that Works, we focus explicitly on the ascertainment of students' prior knowledge. In this respect, the Activity deals with an aspect of formative assessment discussed in Chapter 7 (see section on diagnostic assessment). Assessment strategies for ascertaining students' prior knowledge are often called "eliciting strategies." It is preferable for these to be embedded in a teaching model as one of its early steps (as in the 5E Learning Cycle or the Interactive Approach), but some instructional sequences may not include an eliciting strategy, or sometimes the instructor may need this information prior to planning instruction. In these instances, eliciting strategies are used independently. While you can readily use this Activity in your methods course, your students can easily adapt it for use with elementary students.

In this Activity that Works, we examine a specific instance of an eliciting strategy that uses student drawings to provide windows into their ideas. Your prospective teachers simultaneously undertake the activity as learners and see the teaching of it modeled, enabling them to incorporate it into their own science PCK. In the strategy's simplest form, students are given a sheet of paper and are asked to draw an aspect of the science topic being studied. For instance, at the beginning of a water cycle study, students may be asked to explain where clouds and rain come from using a drawing. A subsequent one-to-one conference about the drawing can readily allow the instructor to identify what the student knows, any gaps in knowledge, and possible misconceptions. Another science topic for which we have found simple drawings followed by teacher conferencing useful, is the insect life cycle: after showing your methods students a butterfly, ask them to draw where it came from (alternatively, if you are using mealworms at some stage—see ATW 3—you could ask them to draw where the beetles came from).

Activity Overview

Inquiry investigations into the human body do not usually lend themselves to the 5E or Interactive Approach teaching models because they are designed for hands-on investigations and direct involvement with materials—not feasible with the human body! In topics such as this, student drawings can be used as an initial step in order to elicit student ideas. In the Activity, methods students first draw an outline of a human body, then draw and label the "things" inside the body. The "things" can be limited to the torso, or can include the whole body. Body parts related to gender will certainly be mentioned and possibly drawn. If desired, the instructor can attempt to avert this by stating that the body is neuter (neither male nor female). However, since the prospective teachers in a methods course are mostly adult females, this should not be an issue—we usually let them choose whether to draw a male or a female. However, you may wish to discuss with your students how they might handle this if using the activity with elementary students.

The learning goals for this eliciting student ideas Activity that Works include helping methods students build understanding of science teaching, science concepts, and the nature of science:

Pedagogical Knowledge Objectives. Learn a number of strategies for eliciting elementary students' prior knowledge, and understand the importance of accessing and developing pedagogy based on what students already know.

Science Content Knowledge Objectives. Develop an awareness of the location of a variety of body organs, and how they fit within the different bodily systems (the objective is most appropriate for the lower elementary grades)

Nature of Science Objectives. Science involves human construction. For example, scientists group or classify aspects of living things (in this case, bodily organs) according to function, as a way of understanding better how they interrelate. Classification schemes do not arise from nature itself, but are imposed by scientists.

To implement the eliciting strategy in your methods course, you will need to consider the following:

- *Timeline.* The activity itself takes about 45 to 60 minutes. You may prefer to follow this Activity with an inquiry investigation into the human body (recommended), in which case you would require two to three sessions. It is best scheduled after using the 5E Learning Cycle or Interactive Approach Activities that Work (1 and 2), or both.

- *Materials.* Felt pens, adhesive (for fastening sheets of paper to a wall), and large sheets of paper are needed. We have found that discarded ends of newsprint rolls (obtained from a newspaper printer) are ideal—cheap, wide, and can be cut to the desired length. Suitable floor space in the methods room or hallway will be needed.

Activity

Eliciting Prospective Teachers' Prior Knowledge of Eliciting Strategies

We commence this activity by reminding the prospective teachers that, in beginning a science unit of instruction, it is important to identify what the students already know about the topic to be investigated. We then invite them to find a partner to work with, and together make a list of strategies that might be used to elicit elementary students' prior knowledge at the beginning of a science unit of work. Their lists should include ideas drawn from strategies already employed or read about in the methods course or those they have been exposed to in their own science learning experience. When they have finished, we draw up a collective list on the whiteboard. The range of strategies varies with the extent of the students' experience and whether the methods course is early or late in the preservice program. The list usually contains strategies like students' questions and possible answers (see ATW 2), brainstorming (see ATW 1), concept maps, KWL charts, and interviews about instances/events. We sometimes find we have to add to the list if some obvious strategies have been omitted (such as interviews about instances/events), and we ensure that drawing is included. Any strategies that need clarification are reviewed. We usually ask the pair that proposed the strategy to explain it, and then provide further clarification if necessary.

The pedagogical principle here is that it is important to elicit students' prior knowledge before a teaching sequence, and base the pedagogy on the ideas elicited. Most prospective teachers will probably not recognize that this introduction is an eliciting strategy in itself (eliciting knowledge of eliciting strategies), so it needs to be explicitly pointed out to them toward the end of this part of the activity. We then provide a brief summary of the activity, highlighting that it is a modeled example of an eliciting strategy—specifically, using drawings to elicit students' ideas.

Modeling the Eliciting Strategy

We organize the prospective teachers into groups of three, and ask them to self-assign group roles of equipment manager, reporter, and group director (we find that prospective teachers need to work with group roles consistently to be ready to use this routine in their own teaching). Sometimes a group of four has to be formed, in which case a new role can be invented. For instance, the reporter role can be divided into reporter and recorder.

We then introduce the topic of "The Human Body" for the investigation, and give instructions for the activity:

1. Equipment managers collect a set of colored felt pens and a sheet of newsprint.
2. The equipment manager lays the newsprint on a clear spot on the floor, and invites the reporter to lie on the sheet of paper, arms flat on the floor, and legs slightly apart. It is not necessary for the sheet to be covered by the lower legs, since the focus is on the torso area.
3. The other group members trace an outline of the reporter's body onto the paper, taking care not to mark the person or any clothing. The pelvic region is not to be traced, but filled in later.
4. After giving the above instructions, we ask the students to commence. We move around quickly to check that everything is being done correctly, and to spot any inappropriate sexual gestures/behavior.
5. Once all groups are finished, and the model is again standing, we ask them to draw some of the "things" that are inside the body torso. They work as a group to try to ensure that each "thing" is drawn in the place where it would usually be found in a real body, and should also be about the right size and shape. Finally, each "thing" should be labeled with its name (scientific/physiological names are not required). Give a hint that they may wish to use different colors for different "things." We usually give them a time limit of 30 to 40 minutes for this part of the Activity.
6. We continually check on each group's progress, ensuring the students are on task. We make no judgmental comments about what is drawn, even if asked whether it is correct. We do ask for group members to explain aspects of what they have drawn, but are again careful to neither affirm nor deny the accuracy of what they say. If pressed, we say something like, "We'll be working on clarifying some of those things later."
7. After a 2-minute warning that time is running out, we ask them to complete the "thing" they are working on, and then display their drawings on the wall.
8. The prospective teachers are then invited to inspect the other groups' drawings, and note any similarities and differences. A variation we sometimes use is to have each group briefly describe their members' drawing, but this does not work well for larger classes, as the time taken can be problematic for schedules and retaining students' attention. Another variation is a gallery walk, where each group visits each drawing for a few minutes, writing comments/questions on a sticky note for the authors/artists to consider.

The completed drawings and conversations with students while they worked reveal the extent of their knowledge of the main organs in the human body. We usually find they have vague ideas about the size and location of most organs, especially the kidneys and liver; and frequently do not include organs like the pancreas. Occasionally one group will produce a fairly accurate drawing, usually because a group member has previously studied human biology.

To conclude the examination of eliciting strategies, we ask our students to investigate other eliciting strategies and develop brief lesson notes for using one of them to introduce a science topic. In particular, we ask them to show how they would use their knowledge of students' prior ideas as a basis for their subsequent pedagogy.

As recommended earlier, we consider it desirable to use this eliciting strategy as the first step in an inquiry investigation undertaken by our prospective teachers. Our prospective teachers often recognize the desirability of eliciting students' prior knowledge, but fail to use this information in their subsequent teaching; so explicit modeling of this helps them build this into their science PCK.

Modeling a Follow-up Inquiry Investigation

We find that our methods students are usually keen to find out the extent to which their drawings are accurate, so this is a natural next step. This usually requires some book or Internet research, but can also include examination of a plastic model of a human torso, if one is available. This subsequent investigation can be fairly simple (as would be appropriate at the Kindergarten level), where students complete a class drawing (or even better, a collage) by putting the major organs in their respective places. Alternatively, the investigation can be complex (as would be appropriate to grade 4 students), by identifying and grouping organs according to their function as classified by scientists (e.g., the blood/heart/lung system, the food/gastric system). This provides an excellent opportunity for talking about the role of human construction in science. You can help the methods students to see that, although scientists could "discover" the arrangement of organs in the human body, they "constructed" the idea of various body systems and which organs belong with which systems.

A final step in this investigation can be a repetition of the eliciting strategy used as a concluding assessment tool. That is, methods students are again asked to draw and label "things" in a drawn torso. If done as a group exercise, the accuracy of the drawings and labels provides assessment data about the efficacy of the teaching and learning. If data about individual student performance were desired, it would be necessary for each student to complete a drawing, preferably followed by an interview where the student explained the drawing.

Pedagogical Principles Embedded in this Activity that Works

- When eliciting students' prior knowledge, it is essential that the instructor does not provide clues about the correctness or otherwise of the ideas being elicited. This is because students who are fearful of science may not contribute for fear of being wrong (again), and such judgments tend to inhibit the contribution of further ideas.
- The elicitation of students' prior knowledge is an essential prerequisite for planning an inquiry investigation suitable for the students.
- The process of eliciting students' prior knowledge often results in the students becoming aware of gaps in their own knowledge, with a resultant desire to investigate further. This desire can be used to motivate students to undertake the remainder of an inquiry investigation.
- Comparing one's own drawings with those of others is a useful means of helping students confirm the possible accuracy (or otherwise) of their ideas, and of providing some alternative ideas for further investigation (e.g., "Are the kidneys really that high in the torso?").
- Managing small group work is made easier for the instructor when individuals in a group each have their own defined role. The optimum group size for most science investigations is three, but could range from two to four depending on the age of the students and the science investigation.

Making the Activity Work for You

Much of the research into students' prior knowledge in science has been conducted using one-on-one interviews, such as those used by Osborne and colleagues (Biddulph & Osborne, 1984; Osborne & Freyberg, 1985). Novices familiar with this research can readily conclude, on the one hand, that conducting interviews is the only way of accessing prior knowledge, and, on the other hand, that one-on-one interviews are impractical because of the time and organization required in a busy classroom. Methods students consequently tend to jump to the conclusion that eliciting prior knowledge is usually not feasible in a classroom setting. However, the strategy of using student drawings as both an initial activity in a unit of work, and as a means of accessing students' prior knowledge, provides prospective teachers with another way of finding out about students' prior knowledge not dealt with in other parts of the methods course.

You can also use drawing to elicit your prospective teachers' views about teaching science by using the Draw-A-Science-Teacher Test (DASTT) (Thomas, Pederson, & Finson, 2001; see Chapter 4). In this exercise, you provide your students with a sheet of paper and ask them to draw themselves teaching a science lesson, adding any necessary labels and explanations about what the students and the teacher are each doing. Thomas et al. provide an effective means of scoring the test that is most useful in research contexts; but even without using their rubric, just examining the drawings can provide valuable insights into your students' views.

This ATW is best used in your methods course in association with other components of the course. For instance, an assignment we have used is to have our methods students (in small groups) prepare a unit of work extending over several lessons that they will implement with a group of elementary students. Two of the requirements for the unit of work are that it commences with an eliciting strategy, and that the subsequent pedagogy be built on the information about students' prior knowledge acquired from the eliciting strategy.

In summary, we have found that methods students' drawings provide effective windows into what they already know about selected science topics. The drawings can be done as a group (as in this Activity), or individually. If done individually, it is wise to follow up with a brief conference with the students, so they can explain aspects of their drawing and so the instructor can probe ideas further.

We have found that prospective teachers do not always understand how to use what they have learned about the students' prior knowledge to build a suitable learning sequence. They will often use a strategy to elicit prior knowledge, then ignore what they have found and proceed with a sequence of loosely associated activities (Moscovici & Nelson, 1998). Even after seeing the teaching models outlined in Activities that Work 1 and 2, many prospective teachers still do not grasp this important aspect of science teaching. Multiple examples, such as those in this Activity, are essential in helping prospective teachers assimilate the appropriate use of prior knowledge into their PCK for teaching elementary science.

More to Explore

There is an excellent set of assessment probes developed by Paige Keeley and associates, published by NSTA Press that are particularly useful as eliciting devices. Providing this resource to your methods students as they develop their own science lessons and units helps them to solidify the importance of eliciting student ideas.

Keeley, P. (2008). *Science formative assessments: 75 practical strategies for linking assessment, instruction, and learning.* Thousand Oaks, CA: Corwin Press and NSTA.

Keeley, P., Eberle, F., & Dorsey, C. (2008). *Uncovering student ideas in science, Vol. 3: Another 25 formative assessment probes.* Washington, DC: National Science Teachers Association (NSTA) Press.

Keeley, P., Eberle, F., Farrin, L., & Olliver, L. (2005). *Uncovering student ideas in science, Vol. 1: 25 formative assessment probes.* Washington, DC: National Science Teachers Association (NSTA) Press.

Keeley, P., Eberle, F., & Tugel, J. (2007). *Uncovering student ideas in science, Vol. 2: 25 more formative assessment probes*. Washington, DC: National Science Teachers Association (NSTA) Press.

Keeley, P., & Tugel, J. (2009). *Uncovering student ideas in science, Vol. 4: 25 new formative assessment probes*. Washington, DC: National Science Teachers Association (NSTA) Press.

References

Biddulph, F., & Osborne, R. (1984). *Making sense of our world*. Hamilton, New Zealand: Waikato University Science Education Research Unit.

Moscovici, H., & Nelson, T. (1998). Shifting from Activitymania to inquiry. *Science and Children, 35*(4), 14–17, 40.

Osborne, R., & Freyberg, P. (1985). *Learning in science: The implications of children's science*. Auckland, New Zealand: Heinemann.

Thomas, J. A., Pederson, J., & Finson, K. (2001). Validating the Draw-A-Science-Teacher-Test Checklist (DASTT-C): Exploring mental models and teacher beliefs. *Journal of Science Teacher Education, 12*, 295–310.

Activities that Work 5

Using Models and Analogies
Electric Circuits

Introduction

In Chapter 1, we discussed a view of learning. Two key components in the view of learning were Seeking Information and Processing Information. When learners are confronted with an apparent gap in their knowledge, they seek relevant information and cognitively process what they find in an attempt to fit new knowledge into their existing knowledge structure, and thus fill the gap. This is also an important part of learning through inquiry. Sometimes, however, this proves difficult if the information they find does not make sense to them. This usually happens when the information is couched in technical terms, or requires other prior knowledge that the learner does not have. In science, these difficulties occur most frequently when dealing with abstract concepts such as light (see ATW 6), molecular and atomic models, magnetism, and electric current.

Scientists often resort to using analogies and models (a particular type of analogy) to represent such abstract concepts; effective science teachers also use analogies and models to help their students make sense of the information they are attempting to process. An analogy provides a way of comparing similarities between two different concepts (Glynn, 1991). One concept is familiar, and constitutes the analogue; the other is the difficult concept that the analogue helps us understand via the mutual similarities. A model is a representation of a phenomenon or concept, often analogical in nature (Gilbert, Boulter, & Elmer, 2000). It is consequently essential that our prospective teachers become aware of the importance and usefulness of analogies and models when teaching science, and build a number of them into their science PCK. We are therefore careful to incorporate specific instances of using analogies and models in our methods courses, and explicitly draw our prospective students' attention to their use. We ensure that we also highlight the context in which they are used and where they are placed within the overall instructional sequence. (ATW 6 demonstrates the use of another type of model, the physical model, in science class.)

Analogies and models for teaching and learning are typically obtained from print/online resources or the instructor, though sometimes the students may devise them. As a general principle they should only be introduced when the students are struggling to make sense of information they have accessed. This would be, for instance, in the Explain step in the 5E teaching model, or in Step 5, Specific Investigations in the Interactive Approach (after the investigations have been conducted) (see Chapter 6).

Activity Overview

For this Activity that Works, the use of analogies and models is embedded in another teaching approach that has three simple steps: Orient (focus on the topic, access prior knowledge), Enhance (a scaffolded sequence of activities leading students to the desired learning), Synthesize (review, reflection, and extension of the ideas learnt). This OES teaching model is another form of structured or Guided Inquiry. While the outline of the teaching model is included to provide context for the use of analogies and models, the main focus of the Activity is the different uses of analogies and models in science and science teaching. Models used in science are included, as well as simulation analogical models for teaching and learning. The main focus is on the latter. We emphasize that the introduction of analogies and models cannot be done in isolation from the pedagogical context, so should be embedded in a teaching model such as OES, 5E, or the Interactive Approach. The science topic for the investigation is electric circuits.

The learning goals for the models/analogies electric circuits include helping methods students build understanding of science teaching, science concepts, and the nature of science:

Pedagogical Content Knowledge Objectives. Understand the importance of using analogies and models as an aid to student learning, and begin to build a repertoire of suitable models/analogies.

Science Content Knowledge Objective. Understand that electric current in a circuit is the movement of charged particles from one atom to the next, so that energy is carried from an energy source (e.g., dry cell) to an energy conversion device (e.g., bulb).

Nature of Science Objective. Recognize that scientists develop analogies and models to explain natural phenomena that they cannot observe directly.

To implement the models/analogies electric circuits investigation in your methods course, you will need to consider the following:

- *Timeline.* An abbreviated version of the Orient step takes about 30 minutes. A brief overview and sampling of initial activities/investigations in the Enhance step takes another 30 to 60 minutes (depending on the time available in the methods course schedule some of these

investigations could be undertaken in more depth, which would mean spreading this step over another session). The teaching of the simulated analogical model requires a further 30 to 45 minutes. We usually schedule this session after modeling the 5E and Interactive Approach teaching models.
- *Materials.* Orient step—flashlight bulbs (preferably 1.5 v), batteries (dry cells), two pieces of insulated wire with insulation removed from ends. There should be sufficient materials for each student. Enhance step—flashlight bulbs,[1] bulb holders, batteries (in holders if necessary), insulated wires with alligator clips fitted on the ends, and two center reading ammeters. For the model—packets of candy, like sugar-coated chocolate, and a small stepping stool about 1 foot high. The furniture may need to be cleared from parts of the methods room to allow the prospective teachers to move in a circle, or the group moved to another suitable space.

Activity

An inquiry investigation into this science topic could readily be done using the 5E teaching model. Because the 5E approach was used in ATW 1, we chose a simpler teaching approach, the OES (see Chapter 6), to allow us to accentuate the use of models in explaining the science ideas related to electric circuits. There are three instances of the use of analogies/models in this ATW: (a) sketches to represent different ideas about movement of electric current in an electrical circuit, (b) stylized sketches to represent electrical circuits (model), and (c) an enacted pedagogical model to aid understanding about electric current and energy in a circuit. A guideline that we emphasize with our prospective teachers is that a model or analogy for enhancing understanding is never introduced near the beginning of a unit of work: it is best introduced after the learners have been investigating the topic for a while, and are grappling with understanding the science ideas that they have encountered. That is, there is a "need to know." We therefore have found it best to demonstrate this for our teachers, creating in them a "need to know" before introducing the model or analogy. Traditional science models such as circuit diagrams can similarly be introduced after the methods students can see for themselves the need for a systematic and universally understood way of communicating. The first few steps in the Activity are therefore included in our description, as these are necessary to create the "need to know."

---

1. LED bulbs cut from strings of low voltage decorative lights can be used. A "holder" is built into the light assembly.

Step One: Orient

We introduce the science topic, electric circuits, and proceed immediately to the introductory lesson. This is designed as a means of engaging the methods students, and as an entrée to accessing their prior knowledge. We ask them to do this activity individually. We distribute to each methods student a flashlight bulb, a battery, and one piece of insulated wire with the insulation removed from the ends.[2] We then ask them to make the bulb light. Some may achieve this quickly, but others will take some time. Once one student manages to make the bulb light, the others will copy what they did, though there could be a few who think it is cheating to copy, and struggle to do it. We usually step in to "check that their bulb is working OK," and quickly make the bulb flash. This is usually sufficient for them to see what we did, and make the bulb light themselves. If they still cannot do it, we ask a neighbor to show them. While the students are trying to make their bulbs light, we move around quickly, taking note of where on the bulb and battery they are touching the wire. We ensure that this activity takes an absolute maximum of 5 minutes.

Once everyone has successfully lit his/her bulb, we distribute a second piece of wire to each methods student, and again ask them to make the bulb light, this time using both wires and without joining the wires together. We introduce the second wire so that later we can show that the electric current flows in both wires—many of our students do not recognize a direct bulb–battery contact is necessary in the first activity as effectively when they use a second wire. It also helps us reinforce the parts of the bulb and battery that have to be connected for the bulb to light. We again observe what the methods students are doing, and help anyone who is taking too long to get it working. This activity should take no longer than 5 minutes.

We then ask the methods students to put the materials aside, and as a class complete two sketches showing how the bulbs were made to light—one for the one-wire instance, and the other for the two-wire instance. As a class, we clarify what parts of the bulb and battery had to be in contact with the wires.

We then ask the students, "When the bulb was lit, was there an electric current in the wires?" Rarely is there anyone who says there was not. We then ask, "Which way do you think the electric current was going in the wires?" We ask several students to explain what they think in terms of each of the one-wire and two-wire instances, being careful not to provide any indication that we agree or disagree with their explanations. Sketches on the whiteboard or overhead projector are helpful. We conclude this step

---

2. Using one wire initially is deliberate, as it highlights a common misconception that electric current flows in only one wire.

Using Models and Analogies 259

with the comment that we will revisit this later. The pedagogical principles employed here are to motivate students to want to find an answer to the posed question, and to find out what prior knowledge students have, as revealed by the ways they attempted to light their bulbs and the explanations given.

Step Two: Enhance

We commence this step by presenting to the students four sketches (see Figure A5.1) representing different models for explanations as to which way the electric current is moving in the wires (Appleton, 1997, p. 53,

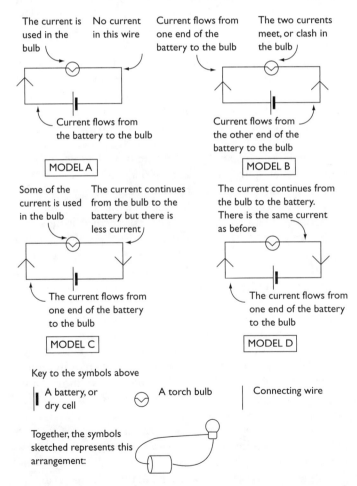

Figure A5.1 Four Electric Circuit Models (source: Appleton, 1997).

Figure 4.1) when one bulb is lit using two wires. These sketches are simplified silhouette drawings of a circuit. The models (i.e., representations of possible science ideas) are introduced as being ideas of students who were asked which way the current was moving. We display each model in turn, and explain how the electric current is considered to be moving in each, as represented by the arrows on the wires. Usually, we find that our prospective students proposed some of the models (or minor variations of them) earlier. With all four models visible, we then ask the students which they think is the best explanation, and take a tally.

We move into the next part of this step by posing the question, "How can we check which of these models is the best?" There follows a discussion of ways of checking the movement of electric current in the wires. If time permits we collate the ideas accepted by the group as valid tests, and use these as the first column in a table (see sample in "Tools" section). The second column is labeled "Evidence Supporting the Idea," and the third, "Evidence Contradicting the Idea." The table is completed as the tests are conducted. Some students may be aware that current can be measured by use of an ammeter, and suggest this. Others will not be familiar with the devices. If no one mentions using ammeters, we introduce the idea. Please note that we move immediately to the most definitive test (using ammeters), because we wish to progress through this part promptly in order to focus on the use of models. However, if you have sufficient time, you could try other tests first. For instance, our students will typically say, "We can try placing another bulb with the first one (in series)—if only the first one lights, that means the current stops there." They can then conduct that test, but ensure that the bulbs are the same voltage and wattage. We also introduce them to a switch, and ask them if it makes a difference where they connect the switch. This convinces them that current is traveling through both wires. These tests clarify that there is an electric current in both wires, but that we do not know in which direction it travels. Introducing the ammeter serves the purpose of showing the current direction.

It is not important that the students know how an ammeter works. We show them an ammeter; pointing out that the needle can move in either direction, depending on the direction the current is moving. Each model (Figure A5.1) is considered in turn, with the students predicting in which direction an ammeter (needle) placed in each wire would move, and whether the reading in each wire would be the same or different. The predictions are recorded.

We then test each prediction. We do this as a demonstration because of equipment availability and the necessity that the ammeters are wired correctly. Each prediction is noted as confirmed or not, and based on the evidence all agree that Model D provides the best explanation. If the term has not arisen earlier, we introduce the idea of an "electric circuit," an expression consistent with current movement Model D. Some methods students wonder how scientists know that Model D is correct. We discuss how, because the actual flow of electricity cannot be observed directly, scientists rely on the model that "works best" to explain the phenomenon.

We now introduce the second type of model: the schematic representation of a circuit. We show the students the standard representations for each of a battery, bulb, and wires, and explain how the wires are drawn straight to keep the sketch simple. We then ask them to construct a schematic drawing of a simple circuit. We provide a brief explanation of why a schematic representation of a circuit is preferable to our earlier drawings, using a complex circuit schematic as an example. The confusing tangle of freehand-drawn wires compared to the schematic circuit drawing makes the preferred use of stylized sketches self-evident. We ensure that we point out the common conventions used in constructing circuit diagrams, which ensure that there is no misinterpretation of its meaning (for instance, symbols for a dry cell, switch, bulb, and wire crossovers).

Many methods students struggle with the idea that the bulb lights up without a loss of current. Electric current is the only aspect of a circuit that they can visualize moving from the battery to the bulb, so intuitively they think that as current moves from the battery, the supply of current in the battery is depleted and eventually runs out (the battery is flat). That is, they have not connected to the idea that energy is transferred from the battery via the current, to the bulb, where some is transformed into light and heat. To help them address this, we then ask them to use book and/or online resources to find an explanation about why the bulb lights, and the nature of an electric circuit. When they return with their explanations, and share them, our prospective teachers are frequently confused by what they have found, and find it difficult to understand. At this point we get our students to participate in the following simulated drama constituting an explanatory analogical model designed specifically to show energy transfer from the battery to the bulb via a movement of electrons—the current (the familiar concept is eating for energy and body movement, and the difficult concept is electric current transporting energy from battery to bulb):

We move the furniture in the methods classroom to the edges to create a large uncluttered area, or move to another suitable space. We ask the students to form a circle around the edges of the room. A small stool placed in the circle with an adjacent desk represents the bulb (place a dish on the desk). The bulb requires energy for it to light, and the students will use energy to step up and over the stool. A supply of candy in a dish is placed opposite the stool, on a desk next to the line of students. The candy station represents the battery, and the confection the energy provided by the battery. The line of students represents the wires, one from the battery to the bulb, and the other from the bulb to the battery. Each student represents a "bit" of electricity in the wire. We explain that scientists call the "bits" of electricity electrons. We create a small space before the "battery," mentioning that when the wires are not connected, the electrons stay put where they are in the wires. We instruct them to close the gap, take two pieces of candy, eat one, and move around the circle. When they get to the bulb, they surrender the other piece of candy as they step over the stool.

After a few moments, we stop the students, and point out what each action represents. At the battery, the electrons receive some energy that pushes them along the wire. They can only move in the same direction, and when everyone else moves. They carry some energy with them. When they arrive at the bulb, they need energy to get through the bulb wire, so surrender some to it (the stepping over the stool, and surrendering a piece of candy). They have sufficient energy remaining to continue moving back to the battery. When the battery runs out of energy (candy), it is flat and can no longer provide energy for the electrons to move. We repeat the action, and then ask some of the students to explain what the components of the model represent. We also ask them to relate the model to the ideas they found from their book and Internet research.

When the room is returned to its normal configuration, we discuss with the students the limitations of the model. We emphasize that models are *representations* of something, and therefore can represent only some aspects of it, and some aspects will not be accurately represented. For example, apart from the obvious disparity between energy in the battery/ electrons and the candy, the dramatized movement of electrons could convey the idea of electricity moving along a tube or pipe—especially if any students had discovered in a book (or online) the water-in-a-pipe analogy for a circuit. When we raise this, we sometimes find that our students want further clarification of how the electrons move, necessitating the use of another model showing electrons passing from one atom to another along the wire (this model in turn has limitations in representing how scientists understand what is happening, but we rarely need to take this further).

We also emphasize that the main purposes of this pedagogical model are to show that: (1) electric current is a continuous movement of electrons along wires in a circuit; (2) the electrons move only when provided energy from the battery; (3) the electrons must have an uninterrupted pathway from battery back to the battery for the current to flow; and (4) some of the energy imparted to the electrons by the battery is released at the bulb as they pass through it. The model does not show other aspects of current flow and energy transfer such as how the battery supplies the energy (i.e., chemical energy to electrical energy), how the current surrenders some energy to the bulb (i.e., resistance), or how the energy released in the bulb is turned into light and heat (i.e., electrical energy to light and heat energy). Other analogies or models may be needed to explain these. This is also an opportunity to mention how many concepts in science are interrelated: the intuitively simple notion of electric current in a circuit can be expanded to fairly complex ideas about atomic structure, electron energy levels, and energy transfer mechanisms. However, we find it unwise to attempt to deal with these content areas in this session—its main purpose is to demonstrate the use of analogies and models.

The pedagogical purpose of this model is to help our prospective teachers understand: (1) in terms of science concepts, the respective roles played by current and energy in a circuit, specifically providing a contrast to the misconception (e.g., Model C) that current is supplied by the battery and consumed by the bulb until the battery runs out of current (goes flat); and (2) in terms of elementary science pedagogy, that analogies and models can be used to help students understand complex ideas in science, but there are limitations to models that need to be explicitly explored.

Step Three: Synthesize

With the students in pairs, we give each group a few flashlight bulbs in holders, batteries (in holders if necessary), and several wires with alligator clips on the ends. We ask them to first make one bulb light, then two bulbs. We then ask them to light two bulbs again, but this time to ensure that the brightness is different from when they did it last time. The materials are put away, and summary schematic sketches of the ways the bulbs were connected placed on the whiteboard, particularly noting in the two-bulb instance, whether the bulbs were dull or bright.

We introduce the terms parallel and series circuits (if not already used), and invite students to explain why the series circuit bulbs were dull, and the parallel circuit bulbs remained bright. Sometimes they arrive at an acceptable explanation, but we usually find that we need to revisit the drama model. When acting out the model with two bulbs included in series and then parallel, they quickly realize that the series bulbs receive only half the amount of energy each (compared to a single bulb circuit). When modeling a parallel circuit, they see that each bulb receives the same amount of energy, but they have to move more quickly to do this, and the energy supply (of candy) is depleted more rapidly. This provides another opportunity to illustrate the limitations of models: because our students have to move more quickly around a parallel circuit does not mean that electrons increase their speed—more electrons are really moving, but that is hard to replicate in our model.

The pedagogical purpose of this part of the activity has two dimensions: (1) to model aspects of the science pertaining to current movement and energy transfer in series and parallel circuits in order to support explanations and observed evidence; and (2) to again show how to use pedagogical models as a way of helping elementary students access complex ideas in science.

Unless we have plenty of time, we usually describe the subsequent activities, rather than work through them with our methods students. The first of these activities is a problem such as devising the wiring for a dolls house (or similar, depending on gender) so that the light in each room can be switched on or off. The final activity is a written task, using schematic sketches, to show the dolls house wiring and explain the movement of the current.

We conclude the ATW by reviewing the use of models and analogies, and highlighting some other instances when models are useful, in science and in science teaching. For example, in science, the chemist August Kekulé envisioned a snake chasing its tail and recognized the ring structure of benzene, and George de Mestral invented Velcro® after returning from a hunting trip and noticing the burrs sticking to his clothes and his dog's fur. In science teaching, drama models are effective in helping students understand about evaporation, surface tension, dissolution, and crystallization (see "Tools" section). We also emphasize with our students the importance, when using these simulation models in elementary science, of having well established routines and rules for managing the class: there is obvious potential for misbehavior (see next section).

Making the Activity Work for You

Three important aspects of modeling are built into this ATW: (a) the use of models to aid communication of ideas, (b) the use of models to simplify complex situations, and (c) the use of models as a pedagogical tool to aid understanding of abstract ideas. The first two uses of models relate to science and the nature of science, while the last obviously relates to PCK for science teaching. Models are used frequently in science, so it is important for some common examples to be introduced to prospective teachers, because they will need to teach these to their students. Perhaps more importantly, it is essential for prospective teachers to recognize models as a pedagogical strategy to help students understand complex ideas and build the use of models into their repertoire of PCK for instructional strategies. ATW 6 also uses models that serve a dual purpose—representing complex scientific ideas, and providing a pedagogical device to enhance understanding. Please note that we have not included in this or other ATWs any analogies or models using computer simulations. We do not have strong confidence that individual use of computer simulations by elementary students is highly effective, so given the time pressures on our methods class schedule, we have chosen not to include this aspect in our discussion of analogies for science teaching and learning.

We have, however, found that this ATW provides opportunities to draw the attention of our prospective teachers to several routines (see Chapter 6) that are needed for the activity to be conducted successfully with children.

Doing a drama modeling exercise as suggested here requires that there be established routines for giving instructions/directions, obtaining attention even when there is a lot of noise and activity, and managing behavior during non-seat activity. Depending on your methods class schedule and time availability, you may prefer to combine the instruction about models and analogies with instruction about one of the other teaching models, rather than introducing another teaching approach. This can readily be done, provided care is taken to explicitly point out the respective roles of student and instructor, and the pedagogy being used.

It is important when you use analogies or models in science or as a pedagogical tool that you make explicit how the familiar aspect of the analogy/model relates to the science phenomenon, and the aspects of the analogy/model that do not map well onto the science phenomenon. As mentioned earlier, an analogy or model is a representation of some science phenomenon that usually involves complex components and/or ideas. As such, parts of the representation closely resemble the phenomenon, and other parts do not. We need to ensure that our students (and our methods students need to ensure that their elementary students) are aware of how the analogy/model maps onto the phenomenon, and where the relationship breaks down. If we do not do this, there is the real danger that students can develop misconceptions. For instance, there is the well-known example of a grade 2 student being very perplexed after his teacher explained about evaporation and cloud formation using a kettle of boiling water, and condensation forming on a mirror. After gazing out of the window at the clouds, he asked, "Please miss, where's the kettle?"

Tools for Implementing the Activity

Examples of Evidence Tables (See Step 2 of Activity)

Circuit Model A

Test	Evidence supporting the idea	Evidence contradicting the idea
The first "light the bulb" activity with one wire.	There was only one wire in the circuit (i.e., the students do not recognize the direct bulb–battery contact as effectively being a wire).	The two-wire activity works only when the second wire is used.
Insert a second bulb in the circuit (i.e., in series).		Both bulbs light, but are dull.

Circuit Model B

Test	Evidence supporting the idea	Evidence contradicting the idea
Set up a bulb and battery circuit and leave it till the bulb goes out—the current coming from each side of the battery is used up.	The battery goes flat.	There is no direct evidence that the current travels from each end of the battery.
Place a center-reading ammeter on the left of the bulb then on the right. The reading on the left of the bulb should be to the right side of the meter, and the reading from the right of the bulb should be to the left side of the meter. The amount should be the same but in opposite directions.		The ammeter gives identical readings in the same direction.

Circuit Model C

Test	Evidence supporting the idea	Evidence contradicting the idea
Insert a second bulb in the circuit (i.e., in series)—the second bulb should be duller than the first.		Both bulbs are the same brightness (both dull).
Place a center-reading ammeter on the left of the bulb then on the right. One should give a smaller reading, both in the same direction.		The ammeter reads the same amount of current on both sides of the bulb.

Circuit Model D

Test	Evidence supporting the idea	Evidence contradicting the idea
Place a center-reading ammeter on the left of the bulb then on the right. The readings should be in the same direction and identical.	The ammeter reads the same amount of current on both sides of the bulb, in the same direction.	

Examples of Other Dramatizations (see Step 3 of Activity)

Evaporation, Surface Tension, Dissolution, and Crystallization

A common thread in water dramatizations is for a selection of the students to be given a blue ribbon to wear. Each represents a molecule of water. They move slowly around a defined area of the room, hands outstretched so they briefly touch other students' hands. This simulates both the kinetic movement of molecules, and hydrogen–hydrogen bonding (hand touching). We show how dissolving is represented by introducing a clump of (say) sugar into the wandering students. The clump consists of several other students wearing a different colored ribbon, all holding another student's hand firmly. The students in the center of the clump have both hands occupied, but some of those on the periphery have a free hand that is outstretched into the "water." We ask one of the "water" students to grasp a free hand from a student in the clump and gently draw him/her from the clump. The clump now moves around with the water students holding hands for longer periods and often with two water students at once. This is repeated until the clump is "dissolved." A reverse of this process models crystallization.

Surface tension can be modeled by using different groups of students for both water and air (the air students do not touch each other and move more rapidly). The water students at the boundary therefore have to grasp the hand of those next to them and "below" them, representing the boundary surface tension "skin" and the surface tension force inwards. A methods instructor can extend this model to show evaporation: one "water" student is asked to move more rapidly, and to head toward the boundary. After a fleeting touch of boundary water hands, the water student joins the "air" students and moves around with them (condensation can also be modeled in a similar way). We can extend this further by introducing pairs of students (detergent molecules with one hand water-loving, and the other air-living). When some of these students concentrate at the air–water boundary, the weakened water–water hand holding at the boundary is evident. We have extended this by introducing detergent molecule pairs with water-loving and fat-loving hands. In a similar way to dissolution, we demonstrate how clumps of fat are broken up and fat molecules are carried throughout the detergent water.

More to Explore

A quick guide to analogies in elementary science:

Smith, S. R., & Abell, S. K. (2008). Using analogies in elementary science. *Science and Children, 46*(4), 50–51.

Useful tools for teaching with analogies:

Glynn, S. (1996). Teaching with analogies: Building on the science textbook. *Reading Teacher, 49*, 490–491.
Glynn, S. M. (2007). The teaching-with-analogies model. *Science and Children, 44*(8), 52–55.
Treagust, D. F., Harrison, A. G., & Venville, G. J. (1998). Teaching science effectively with analogies: An approach for pre-service and in-service teacher education. *International Journal of Science Education, 9*, 85-1-1.

Examples of teaching science with analogies and models (others are found embedded in curriculum materials or online):

Harrison, A. (2004). Teaching and learning science with analogies. In G. Venville & V. Dawson (Eds.), *The art of teaching science* (pp. 162–177). Crows Nest, Australia: Allen & Unwin.
Koch, J. (2010). *Science stories: Science methods for elementary and middle school teachers* (4th ed.). Belmont, CA: Wadsworth.

Research on science teaching and learning with analogies:

Chiu, M. H., & Lin, J. W. (2005). Promoting fourth graders' conceptual change of their understanding of electric current via multiple analogies. *Journal of Research in Science Teaching, 42*, 429–464.
Jakobson, B., & Wickman, P. O. (2007). Transformation through language use: Children's spontaneous metaphors in elementary school science. *Science Education, 16*, 267–289.
May, D. B., Hammer, D., & Roy, P. (2006). Children's analogical reasoning in a third-grade science discussion. *Science Education, 90*, 316–330.

References

Appleton, K. A. (1997). *Teaching science: Exploring the issues*. Rockhampton, Australia: Central Queensland Press.

Gilbert, J., Boulter, C. J., & Elmer, R. (2000). Positioning models in science education and in design and technology education. In J. Gilbert & C. J. Boulter (Eds.), *Developing models in science education*. Dordrecht, the Netherlands: Kluwer.

Glynn, S. M. (1991). Explaining science concepts: A teaching-with-analogies model. In M. Shawn, S. M. Glynn, R. H. Yeany, & B. K. Britton (Eds.), *The psychology of learning science* (pp. 219–240). Hillsdale, NJ: Lawrence Erlbaum.

Activities that Work 6

Learning about Discourse
Light and Shadows

Introduction

Individuals who enroll in the elementary science methods course are often science phobic, especially in regards to the physical sciences (see Chapter 4). Thus, we believe that they need many experiences with physical science phenomena and many opportunities in the methods course to build science concepts in conjunction with learning about science teaching and learning. Furthermore, we recognize that talking through ideas, whether in the elementary classroom or in the methods classroom, is a key factor in learning with understanding in any subject (Bransford, Brown, & Cocking, 1999). In this Activity that Works, prospective teachers talk through science concepts and ideas about the nature of modeling in science as they study natural phenomena related to light and shadows. In addition, they develop their PCK for teaching science through classroom talk by reading articles, viewing classroom videos, and analyzing their own classroom talk.

Activity Overview

Light and shadows activities provide an opportunity for elementary education majors to learn basic physics principles. Although introductory concepts about light are appropriate to teach in the lower elementary grades (K-3), many methods students will not have had hands-on/minds-on opportunities to learn them with understanding and may feel uncomfortable teaching them. Or, they may believe mistakenly that light phenomena are straightforward to teach and to learn. In this Activity, the methods students engage in a number of exploratory activities with light and shadows, and then work together to build a model of light and shadows based on their observations. They also learn about the importance of dialogic discourse (Mortimer & Scott, 2003) in elementary science and how to recognize and facilitate different kinds of science talk through scientists' meetings and small group explorations.

The learning goals for this light and shadows Activity that Works include helping methods students build understanding of science teaching, science concepts, and the nature of science:

Pedagogical Knowledge Objective. Build appreciation for the importance of classroom discourse and develop strategies, such as the scientists' meeting, for encouraging dialogic discourse.

Science Content Knowledge Objective. Understand that light travels in straight lines and that shadows are formed when an object blocks light. (These science concepts are most appropriate for teaching in the lower elementary grades).

Nature of Science Objective. Understand that science involves the evaluation of models of the natural world informed by evidence. Models can never be proven, although some models work better than others to explain the evidence (Driver & Leach, 1996).

To implement the light and shadows activities in your methods course, you will need to consider the following:

- *Timeline.* Total time needed: 3–4 sessions of 50 minutes each.
- *Materials.* Per pair of students: flashlight, objects for exploring light (transparent, translucent, and opaque), a collection of kitchen tools (spatulas, slotted spoons, cheese grater, etc.), objects for making shadows (wooden cylinder, clear plastic cylinder, wooden block, dolls), large paper for building the light model, prediction sheet.

Activity

As the prospective teachers come into class, I greet them and invite them to the scientists' meeting (Reardon, 1993), designated by a circle of chairs, where we will have an introductory science talk (Gallas, 1995) about light and shadows. Once everyone is there, I ask methods students to turn to a neighbor and tell one thing she knows about how light behaves. In the large group, each student shares another student's idea, which I record on the whiteboard. Some ask questions, such as, "Do we have to have light to see?" which I dutifully record without comment. After everyone has contributed, I announce, "Today we are going to begin our investigation of light," while passing out a flashlight and a bag containing objects that are transparent, translucent, and opaque to every other student. "Work with your partner and with these materials to see what you can find out about light. We'll meet back at the scientists' circle in 10 minutes." So begins a three-class period investigation of light and shadows, classroom discourse, and modeling in the science methods course.

After this initial exploration of light, students return to the scientists' circle to share some observations—how the light behaves in relation to the objects in their bag, what happens when you move the flashlight in different ways, that light "bounces off" of some shiny surfaces, that they cannot see without light. Some begin to generalize about kinds of objects that light will "go through" and kinds of objects that will "block" the light. At this point, I introduce some new objects—a collection of kitchen tools such as spatulas, slotted spoons, and cheese grater. "How do you think light will behave with these objects?" This time, I also pass out recording sheets with three columns: "Object, What I think will happen (prediction), What happened" (see "Tools" section). The pairs return to their exploration spots to work with the new materials.

After about 10 minutes of exploring, students gather again for a scientists' meeting. As each pair reports on their explorations, we examine our list of light ideas and see if we want to change or add any ideas. The word "shadow" is mentioned frequently as the students talk about how the light interacted with the kitchen tools. "During our next class, we'll explore some more with light and shadows," I promise the students.

To end this class period, I ask students, who have been playing the role of elementary science students, to return to their regular places in the room to debrief the light and shadows activity as future teachers. We start a large chart in the front of the room with two columns—What the teacher did, and What the students did—and we work through the chart together. As the students describe what I did as the teacher, I probe for more information, "Why do you think I started that way? Why do you think I included those objects in your baggies?" "Why did I have you switch locations between the scientists' meeting and the dyad explorations?" When we discuss the role of the student, I ask them how they felt about talking about their ideas during the scientists' meeting. I encourage them to think about how their talk and participation changed between large group meeting and small group exploration. For homework, I ask prospective teachers to read two articles about science talk (Gagnon & Abell, 2007; Reardon, 1993) and come ready to discuss the role of talk in science instruction during the next class.

The next class session, the methods students go directly to the scientists' circle and begin talking about their thoughts about light and shadow since the last class. "I wonder what would happen if we shined the flashlight on foil that was covered with wax paper," muses one of the students. I am encouraged that they have been thinking about light since our last class. Once everyone arrives, I initiate a discussion. "If you had to explain a shadow to someone, what would you say?" Students offer various definitions of shadows, comparing their ideas with each other: "Shadows mean there is no light." "Shadows can be different shades and in different places." "Not all objects will make a shadow."

"In order to clarify our ideas about light and shadows, I have just a few objects for you to explore today." I hand out three objects to each group—a wooden cylinder, a clear plastic cylinder, and a wooden cube—along with big sheets of drawing paper. I ask students to figure out how many different shadows they can make with each object, and trace the different shadows for each object on the same piece of paper.

Student dyads set to work changing the position of the light and the object to create shadows of different shapes and sizes. Some are surprised that they can make a rectangular shadow with both the wooden cylinder and the cube. Others find it interesting that the plastic cylinder, although transparent, makes a shadow. They wonder if our description of shadows will need to be revised. As the instructor, I circulate among the dyads, mainly listening to their conversations and taking notes about the nature of the discourse that I will use in our debriefing later.

Back in the scientists' meeting, I ask each group to share one of their shadow drawings while the rest of the group guesses which object made the shadow. We also revisit their descriptions of shadows, trying to align their observations with their explanations. "In our next class session, we will build a model of light and shadows that may help us with our shadows definition," I remark.

As is typical in our class, I ask the methods students to switch from learning science to thinking like science teachers. First I present two transcripts of classroom talk, one that represents a typical IRE (initiate-respond-evaluate) sequence (Lemke, 1990), and another that is dialogic and interactive (Mortimer & Scott, 2003) in nature. "What do you think about these two science classes? Which one sounds most like the scientists' meeting we just had? Which one do you think would promote deeper science understanding? Why do you think so?" I also share snippets of the conversations that I noted during their earlier shadows explorations, and ask them to think about which of the transcripts their talk resembles. Students use ideas from what they read as homework to discuss the value of talk in the science classroom. For the next class, I assign students to watch a video about properties of light that includes a visit to a second grade classroom where students explore light with activities similar to the ones we just completed (Harvard-Smithsonian Center for Astrophysics, 1999). I ask the prospective teachers to listen to differences in the talk between the scientists' meeting and the small group explorations in the second grade class. I also ask them to remember our past discussions about children's science ideas and listen to these second graders' ideas about light and shadows.

In the third and final class devoted to the light and shadows lesson, we again meet in the scientists' circle. I bring scissors, tape, and large roles of colored paper.

> Scientists often build models to help them explain the observations they make. Sometimes these are drawings, sometimes they are physical models, and sometimes they are mathematical models. I suggest that we work together to create a physical model of what we think happens with light and shadows.

I show them a giant flashlight I cut out of colored paper. "What else do we need for our model?" I ask. Students mention that we will need a screen and an object for the light to shine on; one student suggests we make a cube. I assign a small group of students to build a large brown paper cube. While they work, other students mention that we also need to represent light. "How will we do that?" I ask. Students have an interesting discussion about light. They claim that, because light travels in straight lines out from the flashlight, we should cut out strips of yellow paper to represent light rays. Another student mentions that we have to show the rays of light spreading out from the flashlight, like in their explorations.[1] "We also need to show light and dark places on the screen" (referring to the shadow and shadowless areas). In this manner we build a physical model, taking into account the evidence from 2 days of exploration.

When the model is complete, I wonder aloud how we will know if the model is correct. Some students suggest checking on the Internet. Another says that all they need is their own evidence. "How do scientists know when their models are correct?" someone asks. Eventually we decide that "correct" is not the right word. Instead, we should find out if our model "works" to explain the evidence. We proceed to check our model against the evidence from our light and shadows explorations. Before we finish, I mention that, although physicists have different models of light that help them in their work, our model of light works for elementary students' rudimentary observations of light and shadows.

Back at their science methods tables, students once again switch from wearing their science student hats to their science teacher hats. I first ask them to think about the model-building activity and how it relates to our goals for elementary science. Second, we return to the discussion of science classroom talk. I ask them about the video they watched with the second grade science students.

> How was the talk in the scientists' meeting similar to and different from the talk during small group explorations? Did you notice if the talk was dialogic or not? What evidence is there that second grade students can talk science? What is the teacher's role?

We also discuss what might come next in the learning sequence for these second grade students, or for themselves. We examine science standards and curriculum materials to see what light concepts typically are taught in the lower elementary grades and in the upper grades. To reach closure on the light and shadows/classroom discourse activity, I ask the students to write a minute paper in which they list something they learned about light

---

1. This ray model of light, although not scientifically complete, is appropriate for elementary science.

and shadows, about science teaching and learning, and about the nature of science during the last three class periods.

Making the Activity Work for You

The light and shadows Activity that Works can be undertaken over a few class periods with minimal equipment requirements. Methods students enjoy the explorations and are fascinated by the model-building activity. Furthermore, seeing second graders engage with similar activities helps prospective teachers visualize how the activities would work in an elementary classroom. They gain confidence with the science content and with age-appropriate materials and activities. They begin to understand the role of models in scientific reasoning.

Students also begin to understand the importance of talk in the elementary science classroom, and that talk can vary across settings. During the light and shadows lessons, the methods students engage in scientists' meetings where they share their observations and generate explanations with their classmates. In hands-on activities, they work with a partner, talking about their explorations and offering tentative explanations. With the help of classroom transcripts and video clips, they see different patterns of discourse that can occur in an elementary science classroom, and think about the role of the teacher in supporting classroom talk. Their own experiences with the light and shadows activities make their readings come alive.

For the light and shadows activity to "work," we believe that all three learning goals (related to discourse, light concepts, and modeling) must be addressed through the doing and debriefing process, with methods students consciously switching roles between science learner and science teacher. In this way, prospective teachers get in touch with their own science learning and think about how the various explorations and scientists' meetings might help them facilitate children's learning.

Tools for Implementing the Activity

Light and Shadows Recording Sheet

Researchers: _____

Examine each object in your baggie. For each object:

1. Write the name of the object in the first column.
2. Predict what you think will happen when you shine a light on the object, and write your dyad's prediction in the second column.
3. Try it! Explore various ways to shine the flashlight on the object. In the third column, record all of the things you observe about the object and the light.

Object	What we think will happen	What happened

More to Explore

Barrow, L. H. (2007). Bringing light onto shadows. *Science and Children, 44*(9), 43–45.
Gagnon, M. J., & Abell, S. K. (2008). Explaining science. *Science and Children, 45*(5), 62–63.
Hayton, M. (1995). Talking it through: Young children thinking science. In S. Atkinson & M. Fleer (Eds.), *Science with reason* (pp. 32–41). Portsmouth, NH: Heinemann.
Muhill, D., Jones, S., & Hopper, R. (2006). *Talking, listening, learning: Effective talk in the primary classroom.* London: Open University Press.
Newton, D. P. (2002). *Talking sense in science: Helping children understand through talk.* London: Routledge.
Reddy, M., Jacobs, P., McCrohon, C., & Herrenkohl, L. R. (1998). Science talk. In M. Reddy, P. Jacobs, C. McCrohon, & L. R. Herrenkohl (Eds.), *Creating scientific communities in the elementary classroom* (pp. 57–87). Portsmouth, NH: Heinemann.
Shapiro, B. (1994). *What children bring to light: A constructivist perspective on children's learning in science.* New York: Teachers College Press.
Wellington, J., & Osborne, J. (2001). *Language and literacy in science education.* Philadelphia: Open University Press.

References

Bransford, J. D., Brown, A. L., & Cocking, R. R. (Eds.). (1999). *How people learn: Brain, mind, experience, and school.* Washington, DC: National Academy Press.
Driver, R., & Leach, J. (1996). *Young people's images of science.* Philadelphia: Open University Press.
Gagnon, M. J., & Abell, S. K. (2007). Making time for science talk. *Science and Children, 44*(8), 66–67.
Gallas, K. (1995). *Talking their way into science.* New York: Teachers College Press.
Harvard-Smithsonian Center for Astrophysics. (1999). *Science in focus: Shedding light on science (Workshop 1, Shine and shadow).* Cambridge, MA: Author. Retrieved July 1, 2009, from http://www.learner.org/index.html.
Lemke, J. (1990). *Talking science: Language, learning, and values.* Norwood, NJ: Ablex Publishing Corporation.
Mortimer, E. F., & Scott, P. H. (2003). *Meaning making in secondary science classrooms.* Philadelphia: Open University Press.
Reardon, J. (1993). Developing a community of scientists. In W. Saul (Ed.), *Science workshop: A whole language approach* (pp. 631–645). Portsmouth, NH: Heinemann.

Activities that Work 7

Integrating Language Arts and Science
A Journey through the Water Cycle

Introduction

As discussed in Chapter 4, prospective elementary teachers often begin their methods courses with apprehension or apathy toward science teaching, as well as weak content knowledge. Research shows that this trend carries into teaching; fewer than 3 in 10 consider themselves "well qualified" to teach science (Weiss, Banilower, McMahon, & Smith, 2001). Many of our methods students do, however, feel comfortable with teaching language arts. Similarly, a majority of elementary classroom teachers (77%) consider themselves "well qualified" to teach language arts. Focusing on integrating language arts thus can be a way to build on prospective teachers' strengths, and to help them build a pedagogical repertoire for science. Just as tradebooks may be a less intimidating way for elementary students to explore topics about which they might be fearful (Pringle & Lamme, 2005), using tradebooks in the elementary science methods course may be a less intimidating way for elementary teachers to approach teaching science.

In this era of accountability, many schools are focusing more heavily on subjects that are tested, such as language arts and mathematics. A benefit of integrating language arts and science is that it can prepare teachers to meet standards for both content areas, and allow each subject to receive proper attention without sacrificing the other (Akerson, 2001). Furthermore, tradebooks can expand the possibilities of the classroom by introducing children to new people, exposing children to faraway places and things, "freezing" moments in time to allow for closer observation, and providing a look at processes that take place over distances and/or time scales that could not be easily observed in the classroom (Abell, 2008).

In order to effectively use tradebooks to teach science, prospective teachers not only need to be familiar with a wide range of genres (e.g., biographies, fiction, nonfiction), but also need to know how to choose appropriate books for science instruction. Tradebooks may not contain accurate information, and even when they do they may nonetheless contribute to formation of misconceptions (Rice, Dudley, & Williams,

2001). Teaching prospective teachers to critically evaluate tradebooks is important to their successful use of tradebooks for science teaching.

Activity Overview

In this chapter, we describe an Activity that Works in our methods course to model an Activity that Works with elementary students, and to introduce students to the integration of science and language arts. We use a tradebook initially to help methods students think critically about their own ideas, but also guide them to consider the information included in (or omitted from) the tradebook, or to practice critical literacy. During the lesson, we also focus on the role of drawing (also see ATW 4) in communicating understanding of science concepts related to the water cycle, and the ways in which this scientific model is useful in visualizing the complex journey water takes through our world.

The learning goals for this water cycle Activity that Works include helping methods students build understanding of science teaching, science concepts, and the nature of science:

Pedagogical Knowledge Objectives. Identify ways that children's literature can be used to introduce children to new scientific concepts; recognize the role of graphic/pictorial communication in children's science learning; understand how multiple sources of information and learning experiences contribute to development of students' understanding.

Science Content Knowledge Objectives. Understand the processes involved in transporting water through our world; recognize how human activity plays a part in the water cycle; understand that the water cycle is not linear (common misconception) but rather a model of the multiple pathways water can take through the environment.

Nature of Science Objectives. Understand that scientific ideas are tentative; that is, subject to change with new evidence or new interpretation of existing evidence.

To implement the water cycle Activity in your methods course, you will need to consider the following:

- *Timeline.* Total time needed: three 45-minute sessions, depending on how often and for how long your methods course meets.
- *Materials.* The only materials needed are those provided in the *Project Learning Tree* curriculum guide or created by the instructor—namely, the chance cards and signs for each of the water cycle stations. Other than that, paper and colored markers or pencils are the only necessities.

The Activity utilizes a nonfiction tradebook, *The Water's Journey* (Schmid, 1989) to help methods students critically examine their ideas about the water cycle. Other similar books that detail the processes through which water moves around the earth could be utilized as well, such as *Where does it come from?*[1] (Pearson, 2005). In this Activity that Works, unlike many in this book, we do not rely on hands-on exploration of materials. Rather, following the read-aloud, we conduct a simulation in which students role-play a drop of water traveling through the water cycle (from the *Project learning tree pre K-8 environmental education activity guide*, American Forest Foundation, 1993). While it is often preferable to teach science through hands-on investigation, this particular example illustrates an alternative way to introducing the concept of the water cycle. In the elementary classroom, it would be followed by firsthand experiences with the processes of evaporation, transpiration, etc. involved in the cycle. Table A7.1 provides an overview of the Activity, described in the sections that

Table A7.1 Lesson Overview

Think-pair-share	The instructor begins a discussion about what comes to mind when methods students hear the word "cycle." The instructor then elicits methods students' prior knowledge about the water cycle by asking them, individually, to draw a picture that represents what they know about the water cycle. Each methods student then pairs up with another student to compare drawings, using a different color marker to indicate any new ideas gained from talking with a peer.
Read aloud	The instructor reads aloud *The Water's Journey* by Eleanor Schmid (1989). The book describes the path that water takes through the environment. The instructor asks methods students to keep track of the water's journey as he/she reads, and to add any new information to their drawings using a different color marker.
Role playing	Methods students role-play a drop of water traveling along the water cycle in the game of chance "Go to the Head of the Cloud" (*Project Learning Tree*). Methods students keep a written record of each part of the water cycle they travel through. They make a new drawing and compare/contrast their previous drawings to note how their ideas have changed.

1. Instructors in Australia should consider using a version of this ATW in conjunction with the *Primary Connections*, Year 3 curriculum, *Water Works*. It would fit well in the Explain phase in Lesson 5, using one of the recommended trade books such as *Where Does it Come From?* (Pearson, 2005). The ATW could be adapted to restrict it to water usage, in line with the focus of *Water Works*; or using a different tradebook it could be used to extend *Water Works* to the wider consideration of the water cycle, as outlined here. The latter may be particularly useful in showing prospective teachers how to use a curriculum like *Primary Connections* as a springboard into further investigations.

follow. In each section, we highlight the ways in which we ask our methods students to both experience the activities as a learner, and debrief those activities from the perspective of teachers.

Activity

Part One: Think-Pair-Share

Taking the Role of Science Learners

We begin the Activity with prospective teachers in the role of learners, and we ask them what comes to mind when they hear the word "cycle." Inevitably, among examples of "bicycle," "process," "life cycle," etc., a student will usually propose "water cycle." Following this brief introduction, we ask students, using a marker color of their choice, to draw their conception of the water cycle. Most of the methods students we work with are familiar with the water cycle, and can recite various processes involved (precipitation, evaporation, condensation, etc.). However, we also find that their conception of the water cycle is often simplistic, in that they believe these processes to occur in a linear fashion. This misconception may arise from popular diagrams that depict the water cycle using a series of arrows arranged in a circle. Indeed, this is how many (though not all) of our students represent the water cycle.

After each student has had the opportunity to complete his/her drawing, we ask them to pair up with at least two different classmates to discuss and compare their pictures. Using a different color marker, students make additions and/or revisions to their picture based on what they learned from their classmates' drawings and explanations. When the methods students return to their seats, we ask them to give examples of new ideas and things they learned from sharing their ideas with their classmates. We make the change in their ideas an explicit part of the discussion in this stage of the Activity, in order to prepare for later discussion of the nature of science, and the way in which scientific ideas change.

Thinking Like Teachers

Following this, we ask our students to reflect on how the think-pair-share process supported their learning, and the role drawing played in communicating their ideas. By "putting on their teacher hats," students are able to speculate as to how these activities might similarly support their own students' science learning. Our students also note the diversity of ideas among the class—which allows us to highlight the prior knowledge their elementary students will bring with them to each new learning experience, and the importance of eliciting students' ideas. We transition from this first activity to the next by indicating that, just as they compared their ideas with each other, it is also important for scientists to check their ideas with other sources of information.

Part Two: A Read Aloud

Taking the Role of Science Learners

We next introduce the book *The Water's Journey* (Schmid, 1989) and tell students that they are going to compare their pictures of the water cycle to the account provided by the author. Students select a third color marker to note new information in their drawings as they listen to the story. Schmid's account includes some often overlooked components of the water cycle, in particular human activities that involve moving water from one place to another, however it provides only a partial account of the water cycle. Following the story, we ask students to note things that they added to their drawings. We also ask them to brainstorm anything they included in their drawing that Schmid did not.

Thinking Like Teachers

We debrief the read-aloud activity by having students consider the tradebook itself, and its value in teaching about the water cycle. Students typically point out that while the information is accurate, it is not complete. Some also note that the way the story "ends" may produce misconceptions about the continuous nature of the water cycle. This highlights the importance of multiple sources of information in developing robust understanding of concepts and phenomena. We transition to the next activity by telling students that we'll be considering how multiple learning experiences can also foster a deeper understanding.

Part Three: Role Playing a Drop of Water's Journey

Taking the Role of Science Learners

The next learning experience is one we have used in our teaching of elementary students and found it to be an Activity that Works for elementary science in that it could easily be accomplished with few materials, was engaging to students, and resulted in understanding of important ideas about the water cycle. Because the water cycle was difficult to observe in a single class session, using a role-playing game allowed us to speed up the process to understand the bigger picture. The activity, Water Wonders, from the *Project learning tree pre K-8 environmental education activity guide* (American Forest Foundation, 1993), invites students to travel along the water cycle through a game of chance—students begin at a station (cloud, ocean, plant, etc.), draw a card, and follow the instructions to proceed to the next station. Their journey includes being evaporated, frozen, consumed, and yes—even urinated! We use the directions for the activity, as well as reproducible cards for each station, included in the *Project learning tree* curriculum guide. However, instructors can easily create their own simulation by identifying particular stops along the journey water takes through the water cycle (e.g., cloud, river, groundwater, etc.) and developing a set of chance cards that represent the range of possible processes and destinations water might next travel through the cycle. Examples are provided in the "Tools" section.

Prospective teachers begin at their assigned station, and keep notes of their journey (including where they were, what happened to them, and where they ended up) through 5–10 cycles of the game (five is sufficient, though if time allows more can be completed). We then ask students to compare their journeys to those of two other people. Did they travel to the same places? In the same order? Did they get to these places in the same ways? Students quickly realize that there is not one set path that water takes through the environment—and that, indeed, there are additional paths they had not considered initially.

The methods students, individually, make a new drawing that synthesizes their ideas about the water cycle, and all the processes and components involved. We then ask them to compare their representations of the water cycle (diagrams) to the way Eleanor Schmid represented the water cycle in her book. Students typically note that Schmid describes one particular journey, rather than the entire water cycle. Their drawings, in contrast, provide a richer description of the possible paths water can travel through the environment. It is here that we introduce the term "model" and ask students to consider how the water cycle serves as a model for what actually occurs in the environment.

Thinking Like Teachers

At the conclusion of the lesson, we ask our students to look through the drawings they made at each step, and to consider how each activity contributed to their growing understanding of the water cycle. Most are often surprised to have learned so much about something they were fairly confident they knew initially. We ask students then to consider how what they did is similar to what scientists do. Would scientists change their ideas about a phenomenon? How? Why? Can you think of any examples in which scientific ideas have changed? Do you see this as a strength or weakness of science? Through the discussion, students begin to see the change in their own ideas as a normal part of scientific thinking—that they continued to develop understanding with new evidence and experiences with the phenomenon, as well as through social interaction with peers.

From a pedagogical perspective, methods students typically see the value in communicating ideas, both orally and pictorially, throughout the activities. Similarly, they point out that using their own drawings as a reference to think critically about the information in the book and their own ideas was an effective variation on the "read-aloud" strategies they had considered in their literacy courses. Usually, there is some debate about whether, as teachers, they should use tradebooks, like *The Water's Journey*, that provide an accurate, but incomplete account of scientific ideas. This provides a fruitful venue for considering how to choose science tradebooks (see suggested class readings in More to Explore), and as an introduction to a class assignment in which prospective teachers develop their own science lessons that utilize a tradebook of their choice. Thus, while it is the end of this particular Activity, it just the beginning of students' learning!

Making the Activity Work for You

This Activity works with prospective teachers, in that it results in differences in understanding that the future teachers themselves recognize. It provides rich opportunities to connect to previous/future course topics including misconceptions, conceptual change, metacognition, and alternatives to "hands-on" science instruction. Few materials are needed, and the time the Activity takes is minimal. In this manner, the Activity models the possibility for teaching science with few resources in a short timeframe.

The Activity we have described here is just one of the myriad of ways in which tradebooks can be incorporated into science lessons. If you do not have experience using tradebooks to teach science to elementary learners, you may find resources such as *Picture Perfect Science* (Ansberry & Morgan, 2004) or the Teaching With Tradebooks column in the NSTA journal *Science and Children* to be useful references.

Tools for Implementing the Activity

Sample Water Cycle Station Cards[2]

Cloud

- You fall as rain and soak into the ground, where you are taken in by the roots of a plant. Go to plant.
- You fall as snow onto a mountain. Go to mountain.
- You fall as rain into the ocean. Go to ocean.

---

2. Based on American Forest Foundation, 1993.

Plant

- You are used in the process of photosynthesis, then transpired by the plant, into the air. Later, you condense into a cloud. Go to cloud.
- You are stored in the tissue of a plant. Remain at plant.
- An animal eats the plant. Go to animal.

Animal

- You are exhaled as water vapor by the animal, and eventually you condense into a cloud. Go to cloud.
- You are urinated by the animal, and soak into the ground. Go to groundwater.

Mountain

- You roll down the mountain and become part of a stream. Go to stream.
- You freeze as a snowcap on the mountain. Stay at mountain.
- You are soaked up by a tree on the side of the mountain. Go to plant.

Groundwater

- You wind up at the surface via a spring, where you are lapped up by a thirsty animal. Go to animal.
- You remain underground. Stay at groundwater.
- You are brought to the surface through a well, and are used to water a lawn. Go to plant.

Stream

- You are lapped up by a thirsty animal. Go to animal.
- You flow along and eventually wind up in the ocean. Go to ocean.
- You evaporate and condense into a cloud. Go to cloud.

Ocean

- You are one of the countless drops of water in the ocean. Remain at ocean.
- You evaporate and condense into a cloud. Go to cloud.
- You are taken in by a kelp plant. Go to plant.

More to Explore

Akerson, V. L., & Young, T. A. (2004). Nonfiction know-how. *Science and Children, 41*(6), 48–51.

Ford, D. J. (2006). Representations of science within children's trade books. *Journal of Research in Science Teaching, 43*, 214–235.

National Science Teachers Association. (2008). *Outstanding science trade books for students K-12*. Retrieved April 19, 2008, from http://www.nsta.org/publications/ostb/.

Owens, C. V. (2003). Nonsense, sense, and science: Misconceptions and illustrated trade books. *Journal of Children's Literature, 29*(1), 55–62.

Rice, D. C. (2002). Using trade books in teaching elementary science: Facts and fallacies. *Reading Teacher, 55*, 552–565.

Royce, C. A., & Wiley, D. A. (2005). The common ground: A rationale for integrating science and reading. *Science and Children, 42*(6), 40–42.

References

Abell, S. K. (2008). Children's literature and the science classroom. *Science and Children, 46*(3), 54–55.

Akerson, V. L. (2001). Teaching science when your principal says "Teach language arts." *Science and Children, 38*(7), 42–47.

American Forest Foundation. (1993). *Project learning tree pre K-8 environmental education activity guide*. Washington, DC: Author.

Ansberry, K. R., & Morgan, A. R. (2004). *Picture perfect science: Using children's books to guide inquiry Grades 3–6*. Arlington, VA: NSTA Press.

Pearson, J. (2005). *Where does it come from?* (Series: Water). Sydney, Australia: Pearson Education.

Pringle, R. M., & Lamme, L. L. (2005). Using picture storybooks to support young children's science learning. *Reading Horizons, 46*, 1–15.

Rice, D. C., Dudley, A. P., & Williams, C. S. (2001). How do you choose science trade books? *Science and Children, 38*(6), 18–22.

Schmid, E. (1989). *The Water's Journey*. New York: North-South Books.

Weiss, I. R., Banilower, E. R., McMahon, K. C., & Smith, P. S. (2001). *Report of the 2000 National Survey of Science and Mathematics Education*. Chapel Hill, NC: Horizon Research, Inc.

Activities that Work 8

Seamless Assessment[1]
The Moon Investigation

Introduction

In Chapter 7, we discussed some of the principles that guide assessment in science education. One of those principles is: assessment should provide opportunities for further learning (i.e., be embedded). In our methods courses, we emulate this principle through our own assessment of methods students' learning. At the same time, one of our goals is to teach prospective teachers important ideas for their own science classroom assessment (see Chapter 5). Thus, in this Activity that Works for the elementary methods course, we model formative and embedded (seamless) assessment techniques within the context of studying phases of the moon. Through this activity, students develop PCK for science assessment as well as building their science content knowledge.

Activity Overview

The moon investigation is an extended activity that takes place within the methods course over a number of weeks. Methods students observe the changing patterns in the phases of the moon. They come to class to share their observations and discuss possible models to explain the lunar phases. Each student's moon notebook is a place for recording observations, noting patterns in the data, reasoning through various explanations, and reflecting on how the activity would work in an elementary classroom.

The learning goals for the moon investigation Activity that Works include helping methods students build understanding of science teaching, science concepts, and the nature of science:

---

1. The content of this chapter has appeared in various forms in other publications, including Abell, George, and Martini (2002); Abell, Martini, and George (2001); Abell and Volkmann (2006); and Volkmann and Abell (2003).

Pedagogical Knowledge Objective. Build a repertoire of strategies to embed assessment throughout a unit of science instruction.

Science Content Knowledge Objective. Understand the patterns in the moon's appearance across a month of observations and develop a model that works to explain those patterns.

Nature of Science Objective. Understand that scientific knowledge is partly the product of human inference and imagination, versus discovery (Abd-El-Khalick, Bell, & Lederman, 1998).

To implement the moon investigation in your methods course, you will need to consider the following:

- *Timeline.* Provide 20 minutes of class time every class meeting for four weeks. (This Activity can take place concurrently with other course activities.)
- *Materials.* Each student needs a moon notebook. Teams of students need a meter stick, ping pong ball, and class light source (e.g., projector).

Activity

During the first week of the methods course, we ask methods students to observe the moon for a month and keep a notebook of their observations, ideas, and reflections on learning. We use a number of strategies to help preservice teachers assess their learning as they build understanding. In particular, we model seamless assessment (Abell & Volkmann, 2006; Volkmann & Abell, 2003), in which different strategies address various purposes for assessment at different points in the moon Activity. These assessment purposes coincide with the different phases of the 5E instructional model (see Chapter 6 and Activity that Works 1) as follows:

- Methods instructor identifies methods students' incoming ideas.
- Instructor determines how students are building understanding.
- Students demonstrate their current understanding.
- Students demonstrate the ability to apply their understanding in new contexts.
- Instructors evaluate what the methods students learned.

Identifying What Methods Students Already Know

We begin the moon investigation by asking students to select one of five answers to the question, "Why do you think the moon appears to be different shapes at different times of the month?" The foils include words and pictures that describe common misconceptions as well as the right answer (Driver, Squires, Rushworth, & Wood-Robinson, 1994):

- Planets cast a shadow on the moon.
- Clouds cover part of the moon.
- The shadow of the Earth falls on the moon.
- The shadow of the sun falls on the moon.
- The position of the Earth, moon, and sun affects phases.

Methods students' answers typically represent all of these options. We display their answers on a histogram so that they can see the variation in the class at this point in the investigation.

We move from this assessment strategy to another way of identifying our students' incoming ideas about the moon—the KWL chart (see Chapter 6). The KWL strategy (Ogle, 1986) is one our students learn in their literacy methods courses and feel comfortable to revisit in the science methods course. At the start of the moon investigation, we ask students to list what they already *K*now about the moon and what they *W*ant to know. In the *K*now column, students often list ideas about rotation and revolution of the Earth and the moon, ideas about moon surface features, and length of a day and a month on Earth (sometimes certain facts are questioned by other students, and we indicate this on the chart). Student questions for the *W*ant to know column are wide-ranging and have included:

- Why can't we see the moon but we can see the stars?
- Does the moon rotate or spin on its axis?
- Why do we see the moon in the day sometimes?
- How big is the moon?
- What does the moon look like during its phases and why?
- What is the temperature on the moon?
- How does the moon affect tides?

During subsequent lessons, students analyze the *W*ant to know column to distinguish between questions that can be answered through observation and discussion and questions that might require the use of outside references. As we raise new questions and find answers, we add to the chart, creating a third column to record what they have *L*earned. The KWL chart is posted in the classroom and used as a reference throughout the moon investigation.

Building Understanding

The moon notebook (see "Tools" at end of chapter) is the main data collection tool that methods students use during the moon investigation, and the only artifact that we evaluate for a grade. The moon notebook assignment requires students to observe daily, take notes on their observations, and try to make sense of the data. They write about what they have observed, noting the day, time, and shape of the moon. At times they mention the weather conditions, their viewing location, and the position of the moon in the sky. Some students include data they have collected from the local news or the Internet. At first, preservice teachers do not have much to say, other than they saw, or did not see, the moon. Later, as the number of observations increases, students have more to say and their notebook entries are more detailed. Student notebook entries thus allow the instructor access to their developing ideas.

As the moon investigation proceeds, we ask methods students to go beyond observation—to describe patterns in the data and make predictions. Through writing and drawing, they discuss patterns in the moon's shape and apparent movement. They ask questions and design tests to further their understanding. Eventually, we request that they invent explanations to account for the data. They invent models that incorporate what they know and what they have observed. As instructors, we find the most engaging parts of their notebooks to be when students try to organize and make sense of the data. This is also a point at which we emphasize our nature of science objective—that science involves human imagination. Our students typically believe that science is about discovering the natural world; they do not consider the important role of invented explanations in science. We ask students to think about the moon investigation in terms of what they can and cannot discover through observations. They come to recognize that they can discover patterns in the moon's shape, rise and set times, and location. However, they cannot discover why these patterns happen. That task involves human invention. This idea about the nature of science is revisited in other course activities.

We also ask the prospective teachers to be metacognitive in their notebooks. Students comment on their own scientific thinking and the development of their understanding over time. They also reflect upon being a teacher. They consider what parts of the moon investigation they would be willing to try with their future students. Students often complain about the extended length of the investigation and question whether children could maintain attention across the 4-week unit. Yet by tracking their learning over time, most come to realize that building difficult concepts takes time and effort.

Student notebooks form the focal point for data collecting, organizing, and theorizing during the moon investigation. However, individual observation and writing must be augmented by interactions with teacher and classmates through science talk (Reddy, Jacobs, McCrohon, & Herrenkohl, 1998) if students are to be successful in developing their science and pedagogical theories. Thus, we encourage students to participate in large and small group discussions and later reflect upon those discussions in writing. These reflections are yet another form of seamless assessment.

Each class period, we spend the first 20–30 minutes working on the moon investigation. On a class moon calendar, we record sightings made since the last class meeting. In their notebooks, students record class data (including days when they personally did not see the moon but others did) as well as facts presented by classmates or the instructor (e.g., the length of time for the Earth's rotation or the moon's revolution). This collaborative effort often uncovers problems in data collection and recording, such as how to draw accurate moon representations and how to judge the moon's position in the sky. Such problems may lead to a mini-lesson about using a compass, a student-generated invention of a standardized way of drawing the moon (e.g., using a penny to represent the full moon), or a discussion of ways to measure the moon's position in the sky (e.g., using a fist measurement or an astrolabe (Smith, 2003)).

Oftentimes, students report observations that conflict with each other. They wonder how one person in class could have seen the moon in a certain location in the sky and someone else in another spot, or how one classmate could have seen a lit crescent on the right side of the moon and someone else saw the left side lit. These discrepancies in the data catalyze discussions about why students might be seeing different things. Their reasons frequently include observer error. Often one of the observers retracts her data in the face of overwhelmingly conflicting data, noting that she must have been wrong. However, the situation is not always so easily resolved. For students who are confused about the phases of the moon, the discrepant data may just add another layer of confusion. They might even believe that the moon could appear so different during the same 12-hour period. However, for students who are beginning to put the pieces together, the discrepant data may lead to an entirely different response. These students devise testable ideas and then invite their classmates to take part in an investigation. For example, many times we have seen students generate the need to systematically collect observations across the day, assigning different individuals to different time slots.

Although it may appear that these large group discussions using the moon calendar are lengthy processes, in reality they sometimes occupy only a small part of a given moon session. The rest of the session is devoted to small group data discussions and problem solving. Often, we simply focus teams on one of the questions that arose naturally during that day's large group discussion. In this way, students are encouraged to refocus, extend, and perhaps bring closure to the large group discussion. Breaking the preservice teachers into small groups allows them more time to sort out their ideas. For instance, in one large group discussion (see Abell, Martini, & George, 2001) students began a debate about whether or not the moon phases were caused by the Earth casting shadows on the moon. This common misconception generated discomfort among some students, therefore the instructor encouraged the students to debate this issue in their teams and to formulate a team theory based on their evidence and understanding of the moon phases. After this small group problem solving session, the teams presented their theories and evidence. Such small group sessions allow students to voice their theories safely, check their ideas against others, listen to their teammates' ideas, develop models, and design tests. Later students return to their notebooks to sort out their thoughts, where instructors can see evidence of idea-building.

Demonstrating Understanding of the Phases

After methods students gather enough data to see patterns emerge and begin to pose tentative explanations of lunar phases, we introduce model-making as a way for them to enhance their explanations. Our definition of models includes two dimensional paper–pencil diagrams, three dimensional tactile representations, human simulations, and computer-generated images and simulations. These models help students visualize celestial objects and manipulate them according to their moon theories.

We encourage the methods students first to employ simple diagrams to describe patterns such as the moon waxing or waning, moving across the sky from east to west, or rising later each day. For instance, we might say, "Observe the moon twice tonight and draw a picture of what and where you see it in the sky." Students refer to and modify their diagrams during class discussions, and find the diagrams to be useful communication tools.

When students wonder about the illumination of the moon, we use a projector to simulate the sun and have students hold small white balls in front of the projector to note how they are illuminated. The students realize that the same side of the ball is lit regardless of its position relative to them, due to light reflected from the projector. They are also able to connect this phenomenon to their current theories.

Similarly, as students question how the Earth, sun, and moon move in relation to one another, we introduce another model. We give the students balls of different sizes to represent the three celestial objects, and ask them to simulate the system. The prospective teachers often choose to make models outside of class, using lamps and flashlights and spherical objects they have around their house, and to write about how their models helped them understand the patterns they observed.

Methods students typically come to understand that the moon is illuminated by light from the sun, which is reflected to the viewer. They discover a pattern of changing moon phases over a month of observations. They might even understand that the relative positions of the Earth, sun, and moon are key. However, most have difficulty putting all the pieces together to explain lunar phases. It is at this point that we introduce another model-making exercise, using a small Styrofoam ball taped to the end of a meter stick (Foster, 1996). The ball represents the moon, the person holding the stick represents the Earth, and again the overhead projector is the sun. The "Earth" can see the images of the moon phases as the moon moves around her. For many students, the moon ball model helps them synthesize the theories they have been struggling with in class and in their notebooks for many weeks.

When students have developed a strong understanding of lunar phases, but are still confused by the apparent motion of the moon, we introduce a final model-building activity. In this model the students themselves represent the Earth, moon, and sun and act out their relative positions and motion. We ask students to return to their moon notebooks to write about their current understanding. Most use words and pictures to describe the Earth–moon–sun system, and some recognize that the moon rise/set times are related to the lunar phases.

At some point, about halfway through the moon investigation, we collect student notebooks and offer a first round of feedback. In the spirit of formative assessment, we do not grade the notebooks. Instead, we evaluate the quality of certain characteristics of their work (e.g., the frequency of observations, the use of class data, the development of patterns and predictions, and the thoughtfulness of their reflections about learning and teaching science). We do not score the notebooks on the accuracy of the scientific ideas; instead we look for evidence that students are working to develop understanding. We provide written comments to engage students in dialogue about their scientific thinking (see Reddy et al., 1998) and provide suggestions for improving their notebooks (see "Tools").

Applying Understandings to New Contexts

Once methods students demonstrate, through their moon notebooks and class discussions, that they have a basic explanation for the phases of the moon, we extend their understanding by challenging them to think about new problems to solve. The main instructional strategy, which also doubles as an embedded assessment, is the moon puzzler. Moon puzzlers are teacher-generated problems designed to scaffold student knowledge construction. The early puzzlers help students focus their attention on two patterns: the changing shape of the moon and its changing position in the sky. We presented a couple of puzzlers here, numbered as they are for the course.

Moon Puzzler #2. Last week I saw the moon and it looked like a banana. This week it looks a lot larger. Are my eyes fooling me? Why could this be happening?

Moon Puzzler #3. Yesterday I saw the moon on my way to school. It was in the southeast sky,[2] slightly above the horizon. Today I was going to school at the same time and I did not see the moon! It was clear out and I looked everywhere, but no moon. Where is the moon?

Later puzzlers ask students to resolve some kind of controversy.

Moon Puzzler #4. Someone in my class saw the moon last night at midnight in the southwest. I saw it at 6 p.m. and I swear it was in the southeast. Who is right? Why do you think so?

Some puzzlers ask students to think like science teachers as part of their problem solving.

Moon Puzzler #6. Below are some excerpts from *Harold and the Purple Crayon*, written and illustrated by Crockett Johnson. Harold indicates that the moon is always in the same place in his window. What do you think about Harold's comments? What would you say to him if you were his science teacher?

The last puzzler in the series introduces an enigmatic new pattern—changes in rise and set times—and allows the students who have developed viable explanations to apply them to a new problem. We have invented the moon puzzlers in response to methods students' questions over many semesters, using a variety of sources, including children's literature (containing both scientifically accurate and inaccurate accounts of the phases of the moon) and moon data excerpted from the newspaper, television, or Internet sources. The moon puzzlers model how science teachers assess student understanding via application. Preservice teachers' answers to the moon puzzlers inform our teaching, letting us know when students need more opportunities to build their understanding.

Determining What Methods Students Know

The moon notebooks, data discussions, and moon puzzler activities serve as both instructional activities and assessment tools. They represent seam-

---

2. Because these moon puzzlers were developed in the Northern Hemisphere, sky directions are specific to that context. For groups of students who develop a robust understanding of phases of the moon from the Northern Hemisphere perspective, we challenge them to think about the moon phases in Australia. This new puzzler is complicated by the fact that many moon calendars on the Internet take a Northern Hemisphere perspective without explicitly stating so. We encourage students to think about what happens as a sky observer moves from north to south on the globe, and to look for moon calendars produced in the Southern Hemisphere (websites for observatories in major Australian cities are helpful). Students who solve this puzzler figure out that, in the Southern Hemisphere, they will face north instead of south to view the moon, the moon will still appear to move across the sky from east to west, but the first and third quarter moons, for example, will look the opposite from how they appear in the Northern Hemisphere.

less assessment that is embedded in instruction. Throughout the moon investigation, we make the instructional and assessment strategies explicit to methods students; in other words, we discuss the strategies we have used, and ask preservice teachers to write their reactions to using these strategies in their own teaching in their notebooks. Thus, the moon notebook serves as a model of formative assessment, where students work on their developing ideas and think about their thinking.

The moon notebooks also include a summative assessment component. The summative assessment in the moon notebook is a reflection paper. At the end of the moon investigation, we ask students to write their current thinking about lunar phases and movement, as well as their current beliefs about teaching about the moon. These reflections include representations of conceptual understanding, responses to the moon investigation as learners, and suggestions for their future classrooms. We expect methods students to integrate relevant course activities and readings into their discussion. We collect and score the completed moon notebooks, including these final reflections, using the same criteria as in the first notebook check (see "Tools"). The scoring guide is based on the work the preservice teachers have done to build their understanding, not on whether or not they have reached a full and accurate understanding of lunar phases.

Nevertheless, students do revisit their incoming views about moon phases and write about the shifts in their thinking in their moon notebooks. Typically all students arrive at a more scientifically accurate understanding of the moon's phases. An example student final reflection demonstrates their newfound understanding:

> Initially ... I had no idea of why the moon phases occurred. Today, I believe that the glow of the moon is caused by the sun's rays ... I have learned that the location of the moon and Earth in relationship to the sun determines the phase of the moon. When the moon is between the Earth and the sun, the backside of the moon is entirely lit, however, we do not see this side of the moon, therefore, we see no moon at all. As the moon revolves around the Earth, more and more of the lit portion is visible from Earth. As the moon continues to travel around the Earth, it hits the full moon phase. This happens when the Earth is directly between the sun and the moon. The full moon means that you can see the whole moon because the sun's light is hitting the moon directly and you can see the whole side of the moon. That's why it looks full. When there is a full moon, it can be seen from about sunset to right before sunrise.

Making the Activity Work for You

The moon investigation requires a time investment on the part of the methods instructor. It also involves a risk to instructors and students, who must agree to examine a natural phenomenon over time and struggle to explain why certain patterns occur. Furthermore, the moon investigation challenges preservice teachers' thinking about science teaching and learning. Seldom have they been involved in a long-term science investigation (see also, Activities that Work 3), and sometimes they are frustrated that we do not just tell them the answers and move on.

Initiating a moon investigation in your methods course requires little in the way of materials. However, there are two major requirements. The first is that you will need to confirm your own understanding of phases of the moon. We recommend that you conduct a mini-study prior to introducing your students to the moon investigation. This will allow you to see the patterns firsthand, and to encounter the kinds of difficulties that your students might face. Second, you need to set aside enough time in your course for the moon investigation to occur. We recommend that the moon investigation takes place over 4 weeks, but only as a portion of each class meeting time. Finding a way to schedule the moon investigation to capitalize on the other components of your course can be a challenge. The writing themes described in the Moon Notebook Assignment in the "Tools" section are one way we help methods students connect the moon investigation to important course topics.

The rewards from the moon investigation are great. Prospective teachers experience the thrill of learning science through exploration and explanation. They develop confidence in their own abilities to think through and solve a scientific problem. They come to understand that to be effective, science instruction must engage students in thinking and writing and talking, not just in hands-on activity. The moon investigation has the potential to transform your methods students' thinking about science education.

Tools for Implementing the Activity

The Moon Notebook Assignment

Purpose

To help you think about teaching science, it is important that you think about learning science. In this assignment you will be engaged as a science learner. You will make observations and invent explanations (construct knowledge) to fit your observations. You are going to become a moon watcher!

Background (from Elementary Science Study, 1968)

In times past people regulated many of their activities according to the recurring patterns of familiar objects in the sky. To early civilizations, astronomy was an intuitive science, as it still is today for many peoples of the world. In our technologically advanced culture, however, children grow up unfamiliar with the rhythms of the sky, because they are not directly dependent on those rhythms to determine the clocks, calendars, and maps of their daily lives. Yet for all of us it is important to see that the world in which we live behaves with observable order in some areas, and that we can find rules to describe what happens and predict what will happen.

Content Standards: Earth and Space Science (National Research Council, 1996)

GRADES K-4

By observing the day and night sky regularly, children will learn to identify sequences of changes and to look for patterns in these changes. As they observe changes, such as the movement of an object's shadow during the course of the day, and the positions of the sun and the moon, they will find patterns in these movements. They can draw the moon's shape for each evening on a calendar and then determine the pattern in the shapes over several weeks.

GRADES 5-8

The understanding that students gain from their observations in grades K-4 provides the motivation and the basis from which they can begin to construct a model that explains the visual and physical relationships among Earth, sun, moon, and the solar system. Direct observation and satellite data allow students to conclude that Earth is a moving, spherical planet, having unique features that distinguish it from other planets in the solar system. From activities with trajectories and orbits and using the Earth–sun–moon system as an example, students can develop the understanding that gravity is a ubiquitous force that holds all parts of the solar system together. Energy from the sun transferred to light and other radiation is the primary energy force for processes on Earth's surface.

By grades 5-8, students have a clear notion about gravity, the shape of the Earth, and the relative positions of the Earth, sun, and the moon. Nevertheless, more than half of the students will not be able to use these models to explain the phases of the moon, and correct explanations for the seasons will be even more difficult to achieve.

Procedure

The moon must always be someplace, but it is not always possible to find. The moon sometimes looks like a ball, sometimes like a banana. Once in a while, I have seen the moon in the daytime. These are random observations about the moon. In the moon investigation, you will collect data more systematically, which will allow you to detect patterns and invent explanations and help make sense of our moon observations.

For this assignment, you will need a notebook (spiral, three ring binder, or composition notebook). You will watch the moon each day for the next month, making an entry in your notebook each time you make an observation. You will use your eyes and your mind to make sense of any patterns you notice and observations you make. Use facing pages of the notebook to keep records as follows:

- Left side: record when and where you observe the moon. In what direction are you looking? Where is it in relation to the horizon? Draw a picture of what the moon looks like and its place in relation to the horizon.
- Right side: reflect upon yourself as a learner, your feelings about watching the moon, tentative explanations that you have for what you are observing, and questions for further study. Include any reactions you have to being a science student in the activity. You will also be asked to include specific responses in your reflections each week about science teaching and about the nature of scientific inquiry.

What to include in a thoughtfully reflective notebook entry:

1. Be sure to include personal reactions to the work of moon watching.
2. Be sure to include detailed observations, patterns you've discerned, predictions, tentative explanations you are developing, and puzzling questions.
3. You may use notes from class data and discussions to supplement your own. If you do this you must identify it as class data.
4. Be sure to include reflections from the point of view of an elementary teacher, based on your experiences and the course readings.
5. Be sure to include a *final reflection* in which you discuss your most current thinking about the moon and teaching and learning about the moon.
6. Above all, don't get discouraged when you don't see the moon. Instead, try to figure out why you are not seeing it, how this would affect an elementary class doing this activity, and ways to remedy that situation. No entry should be limited to *"cloudy, no moon."*

Weekly Writing Themes

Week 1. Put yourself in the position of an elementary student. How do you feel about observing the moon? Is the activity easy or hard? Is this activity motivating or frustrating? Why? As a teacher how would you deal with students' frustration? Why is it important for you to explore the role of science learner?

Week 2. What patterns are you finding in your data? What predictions can you make? What do the state/provincial standards say about studying the moon, about observing and recording data, and about inquiry-based activity? Specifically mention the proficiencies in your reflection.

Week 3. What explanations can you offer for the patterns you see? Include some diagrams in your notebook that represent the Earth/moon/sun relationships as you understand them. What are some ways of representing the moon that you would engage your students with? Why?

Week 4. Several articles in your course packet discuss the roles of science talk in student learning. In your moon notebook, write an entry that details how you and your classmates have engaged in "science talk" when learning about the concept of the phases of the moon. Include references to course readings.

Final Reflection. Discuss your most current understanding of lunar phases and movement, including diagrams to represent your understanding. Also discuss teaching a moon unit in elementary school. At what grade would this be appropriate? Why? What goals would you have? What instructional strategies would you use? Use the readings in your course packet to help think about how to teach the phases of the moon. Which of their ideas would you use? Not use? Why?

Formative and Summative Evaluation

We will collect your notebook in 2 or 3 weeks to provide feedback and give you some input as to your progress. We will also collect them at the end of the moon study. We will be looking for evidence of the following criteria:

- thorough and accurate record keeping
- consistent observing (each day)
- thoughtful reflection (see list above)
  - about yourself as a science learner
  - about science teaching
- integration of course readings, discussions, and class data into your reflections.

Formative Feedback to an Elementary Science Methods Student about her Moon Notebook

Notes to Lauren:

- You have noticed a pattern about the changing shape of the moon and made a prediction. Good.
- Your theory about the moon being in a cycle and only visible at parts of the cycle at certain times of day is interesting. Why do you think that would be? I think your idea about the orbit of the moon is a fruitful one to pursue. What is the Earth doing while the moon is orbiting??
- Think about where the sun, moon, and Earth would be during a full moon.
- I think you must be seeing the moon in the SW, not the NW???
- You need to include more personal reflections about yourself as a learner in this experience.

Moon Notebook Grading Sheet

Name _____

First round evaluation _____ Final Score (30 pts. possible) _____

Data Collection and Interpretation (12 pts.)

_____ Consistency and accuracy of observations (3 pts.)

_____ References to class data and discussions (3 pts.)

_____ Evidence of predictions, patterns, questions, and tentative explanations (6 pts.)

Reflections on Self as Learner and Science Teaching (18 pts.)

_____ Quality and consistency of daily reflections (2 pts.)

_____ Week 1 theme questions (attitudes toward the moon study) (2 pts.)

_____ Week 2 theme questions (science standards) (2 pts.)

_____ Week 3 theme questions (moon representations) (2 pts.)

_____ Week 4 theme questions (science talk) (2 pts.)

_____ References to course readings (2 pts.)

_____ Final entry (current explanations; reflections on teaching and learning) (6 pts.)

More to Explore

Ansberry, K., & Morgan, E. (2008). Teaching through trade books: Moon phases and models. *Science and Children, 46*(1), 20–22.

Chancer, J., & Rester-Zodrow, G. (1997). *Moon journals: Writing, art, and inquiry through focused nature study.* Portsmouth, NH: Heinemann.

Hubbard, L. (2008). Bringing moon phases down to earth. *Science and Children, 46*(1), 40–41.

Young, T., & Guy, M. (2008). The moon's phases and the self shadow. *Science and Children, 46*(1), 30–35.

References

Abd-El-Khalick, F., Bell, R. L., & Lederman, N. G. (1998). The nature of science and instructional practice: Making the unnatural natural. *Science Education, 82,* 417–436.

Abell, S., George, M., & Martini, M. (2002). The moon investigation: Instructional strategies for elementary science methods. *Journal of Science Teacher Education, 13,* 85–100.

Abell, S. K., Martini, M., & George, M. D. (2001). "That's what scientists have to do": Preservice elementary teachers' conceptions of the nature of science during a moon investigation. *International Journal of Science Education, 23,* 1095–1109.

Abell, S. K., & Volkmann, M. J. (2006). *Seamless assessment in science: A guide for elementary and middle school teachers.* Portsmouth, NH: Heinemann.

Driver, R., Squires, A., Rushworth, P., & Wood-Robinson, V. (1994). *Making sense of secondary science: Research into children's ideas.* London: Routledge.

Elementary Science Study. (1968). *Teachers guide for "Where is the moon?"* New York: McGraw-Hill.

Foster, G. (1996). Look to the moon. *Science and Children, 34*(3), 30–33.

National Research Council. (1996). *National science education standards.* Washington, DC: National Academies Press.

Ogle, D. (1986). A teaching model that develops active reading of expository text. *Reading Teacher, 39,* 564–570.

Reddy, M., Jacobs, P., McCrohon, C., & Herrenkohl, L. R. (1998). *Creating scientific communities in the elementary classroom.* Portsmouth, NH: Heinemann.

Smith, W. (2003). Meeting the moon from a global perspective. *Science Scope, 26*(8), 24–28.

Volkmann, M., & Abell, S. (2003). Seamless assessment. *Science and Children, 4*(8), 41–45.

About the Authors

Sandra K. Abell is Curators' Professor of Science Education at the University of Missouri, U.S., where she directs the university's Science Education Center. She taught elementary science as a regular classroom teacher in Iowa, Iceland, and New Mexico, and as a co-teacher with classroom teachers in Indiana. At the university level, she has taught elementary science methods, secondary science methods, and college science teaching methods. She also has worked in professional development settings with elementary, middle, secondary, and college science teachers in the U.S., Australia, Honduras, Portugal, Taiwan, and Thailand. Her research interests focus on teacher learning throughout the career span and across the grade levels. Abell is a past President of the National Association for Research in Science Teaching (NARST) and editor of the 2007 *Handbook of Research on Science Education*.

Ken Appleton is semi-retired as an Adjunct Associate Professor at Central Queensland University, Australia. After 7 years as an elementary teacher, he moved to a university where he has been involved in preservice and inservice program design, and development and delivery of science and design technology methods courses. He has published extensively in international journals in his areas of interest that include elementary science teaching and learning, constructivism, elementary science teacher knowledge, and science teacher professional development.

Deborah L. Hanuscin is an Assistant Professor of Elementary Education at the University of Missouri, where she holds a joint appointment in the Department of Physics and Astronomy, and Department of Learning, Teaching, and Curriculum. She is a former elementary teacher and informal science educator, and has been involved in numerous professional development programs at the elementary and middle school levels. At the university, she has taught early childhood, elementary, and middle/secondary science methods as well as introductory physics courses for education majors. Her research interests include elementary science teacher learning as well as the development of elementary teachers' and students' understanding of the nature of science.

Index

Page numbers in *italic* refer to tables. Page numbers in **bold** refer to figures.

5E model 130–1, 206
Abell, S. K. 179–81
accreditation bodies 39, 47
action research 184
Activities that Work: 5E Learning Cycle (magnetism) *see* magnetism activity; Appleton's notion of 196; application 208–9; card sort sets 209–15; concept analysis 3; design principles 197–9; eliciting student ideas (the human body) *see* human body activity; engage phase 203–4; functions 196; guided and open inquiry (insect study) *see* insect study activity; integrating language arts and science (water cycle) *see* water cycle activity; interactive approach (floating and sinking) *see* floating and sinking activity; learning about discourse (light and shadows) *see* light and shadows activity; sample chart 216; seamless assessment (moon investigation) *see* moon investigation activity; using models and analogies (electric circuits) *see* electric circuits activity
activity-driven orientation 53
"activitymania" 201, 208
add-on field experience 181
analogies and models activity *see* electric circuits activity
Anderson, C. W., & Smith, E. L. 51
Appleton, K. 3, 24, 31, 33–4, 45, 68, 86, 126–7, 171, 195–6, 259
"apprenticeship of observation" 140, 183, 195

Archimedes' Principle, teaching strategies 225–6
assessment: knowledge of 4–5, 40, 74, 138; main types of 139; PCK for 4; principles of effective *see* assessment principles; scoring and grading assignments 152–6
assessment activities: Draw-a-Scientist test 75; lesson planning task 76–7; Science Autobiography 76
assessment principles: authenticity 148–9; backwards design 145–7; embedded learning 147–8; thinking skills development 149–51
assessment scoring: criterion-referenced 152; learner-referenced 152; norm-referenced 152; rubrics 152–6, 159–60, 165, 167, 237, 240, 242–4, 253
assessments: diagnostic 139; embedded (seamless) 147, 149, 168, 287–91, 293, 295, 297, 299, 301; formative *see* formative assessments; principles of effective 144; summative *see* summative assessments
assignment design 157–61
assignments: examples *156*; field-based 175; instructional materials analysis example 162–5; for learning synthesis 150; Moon Notebook Assignment 296–300; scoring and grading 152, *152*–6; Summative Assessment Assignment and Scoring Tool 162, 162–7; teacher-as-researcher 184; teaching portfolio example 165–7; "Teaching Science in the Elementary

assignments *continued*
   School" interview 190–1; "Thinking Like a Teacher" 184
attitudes: documentation of negative 67–8; and the Draw-a-Scientist test 75; importance of fostering positive 73; improving 201; of preservice teachers 66
authentic assessment 149

backwards design 144–5, *146*, 147, 157
behavioural learning theory 17
beliefs 50–4, 56, 58, 66–8, 70, 139–40, 144, 166, 173, 182, 201, 203
beliefs questionnaire 141
Biggs, J. B., & Moore, P. J. 28

card sort activities 51–2, 54, 60, 62, 151, 203–4, 206
card sort sets: light and shadow 210–11; plants and animals 211; rock cycle 211; understanding science 209–10
case-based instruction 179
case knowledge 195, 199
certification 4, 37, 40, 42
classroom, arranging the 95
classroom management 5, 84, 97, 128
classroom transaction patterns 110, 115–16, 224
Claxton, G. 15
collaboration 37–8, 42, 44–6, 71, 154, 186
collaboration of stakeholders in teacher education 38
concept mapping 13, 113–14, 122, 127, 134, 249
confidence 3, 33, 53, 68, 81, 86, 95, 115, 197, 225, 240, 264, 275, 296
constructivist theories of learning 18–23
cooperative learning 53, 61–2, 173, 183, 206, 221
Cosgrove, M., & Osborne, R. 133
course calendar 61
course goals 4, 40, *41*, 52, 81–2, 139–40, 143–5, 150, 157, 161, 165–7; basis of 4; demonstrating progress towards 139–40, 143–4; example *41*; formulating 40; linking assessment to (backwards design), 144–5, *146*, 147, 157; Science Methods Course 81–2; and teaching orientations 52–3
course outline *see* syllabus
course planning: activities that work 86, 86–91; assessment strategies 84; classroom management 84; content selection 82–91; curriculum knowledge 82–3; curriculum materials 94; development challenges 80–1; general pedagogical knowledge 83; knowledge of students 83; organising matrix 87; planning matrix 100; resource location 85; resource selection *see* resources for teaching elementary methods; safety and ethics 97; sample schedules 87–9, 91; science content knowledge 85–6; syllabus component recommendations 98; timing 82
creativity 70
criterion-referenced scoring approach 152
curriculum: knowledge of 72–3; PCK for 4
curriculum components 81
curriculum resources, finding 73, 85

darkling beetles *see* insect study activity
deep processing 28, 34
demonstration lab 180–1
DESE (Department of Elementary and Secondary Education) 40
design principles 185
diagnostic assessment 139–40, *141*, 151–2, 247
direct teaching 117–18
discrepant events 26–7, 29–30, 32, 89, 291
discussion techniques 110
disequilibrium 20
dramatization (role play) 267–8, 280, 282–4
Draw-A-Science-Teacher Test (DASTT) *141*, 252
Draw a Scientist Test (DAST) 75, *141*, 209
Druckman, D., & Bjork, R. A. 172

Earth and Space Science, content standards 297
electric circuit models **259**
electric circuits activity: enhancement

259–63; implementation tools 265–8; orient–enhance–synthesize model 256–7; orientation 258–9; overview 255, 255–7; practical advice 264–5; resources 256; synthesis 263–4; teaching approach 257
elementary science curriculum 82, 87–8, 150
eliciting strategies: modelling 249–51; in practice 252–3
eliciting student ideas activity *see* human body activity
embedded (seamless) assessments 149, 293
expectations, development of appropriate 73–4

field experience: add-on model 180–1; author's experience 186; benefits 57, 172–4; block model 182; challenges and solutions 174–7; contribution 57; demonstration lab model 180; design principles 185; extracurricular model 179–80; Knowles *et al.*'s theoretical framework 182; methodology 56–7; partners model 181; reflection orientation 56–7; research-based vs theoretical support 173; sample feedback form 187; student reflection tools 182–5; virtual model 178–9
floating and sinking activity: exploration 221–2; investigation planning 223–4; investigations 224–6; preparation 220–1; problem solving 227; reflection 227–8; students' questions 222–3
formative assessment 108, 134, 139, 143, 145, 147, 151–2, 218, 253–4, 293, 295
formative assessment probes *141*, 150, 253–4
formative assessment strategies, distinguishing instructional strategies from 148
formative assessment techniques 140, 142–3
Friedrichsen, P., & Dana, T. 60

general pedagogy, definition 107
Goodlad, J. I. 46
graduate selection criteria 81
guided discovery 122

hands-on activities: data collection 121; encouraging 123; and engagement 131; and exploration 131, 134; in guided discovery 122; and instruction-giving techniques 117; in POE strategy 111
Hanuscin, D. 3, 45, 51, 168, 179, 196
Harlen, W. 58, 115, 126, 136, 157–8, 231, 236
human body activity: follow-up investigation 251–2; learning goals 248; overview 247–8; pedagogical principles 252; practical advice 252–3; procedure 249–51

information processing 18, 22, 24
inquiry: guided and open activity *see* insect study activity; teaching science by 83
inquiry-based strategies 44, 117, 121, 150, 153, 173, 177, 233, 238, 299
inquiry model: evaluation 240; expectations 239–40; rationale 238–9
insect study activity: evaluation 240; expectations 239–40; implementation tools 238–45; learning goals 234; overview 233–4; phases 234; practical advice 237–8; procedure 234–7; rationale 238–9
instruction: analogies 28, 112, 118, 255–7, 259, 261–5, 267–9 (*see also* electric circuits activity); asking questions technique 114–16; case-based 123–5; classroom talk 270, 273–5; classroom transaction patterns 110, 115–16, 224; concept mapping 13, 113–14, 122, 127, 134, 191, 249; curriculum integration 87, 90; direct teaching 117–18; equipment distribution 110–11; explaining an idea 112–13; inquiry 107–8 (*see also* inquiry-based strategies); investigation 118, 119, 119–22, 123, 129, 233; KWL 122–3; learning cycle 42–3, 53, 57, 62, 72, 130, 143, 173, 197, 201–9, 211, 213, 215–17, 247–8; methods students' familiarity with 72; planning process modelling 133–4; practical applications 111–12; questioning 28, 53, 62, 104, 111–12, 115, 173, 206;

instruction *continued*
  routines 83, 102, 108–11, 118–19, 124–5, 135, 173, 221, 228, 250, 264–5; techniques and strategies 111–12; techniques for giving instructions 116–17; types of 108; videocases 55–8, 123, 136, 175, 179; wait time 89, 110–11, 115–16
instructional strategies, PCK for 4, 233, 264
INTASC (Interstate New Teacher Assessment and Support Consortium) 40
interactive approach: implementation tools 229–31; key pedagogical ideas 218; origins and literature 126
interactive approach activity *see* floating and sinking activity
investigations, equipment and materials for 95

knowledge for teaching science teachers 5
Knowles, J. G. *et al.* 182
KWL 122, 249, 289

learner-referenced scoring approach 152
learning: behavioural learning theory 17; constructivist theories 18–23; information processing 18, 22, 24; rote 25, 27–8, 69, 105; schema theory 13–16; situated 172–4; social cognitive theory 17–18
learning about discourse activity *see* light and shadows activity
learning cycle 42–3, 53, 57, 62, 72, 130, 143, 173, 197, 201–9, 211, 213, 215–17, 247–8
learning experiences, adding value to student 71
lesson planning assignment 76–7
lesson preparation method 141
lesson sequence card sort 141
light and shadows activity: implementation tools 276; learning goals 271; overview 270–1; practical advice 275; procedure 271, 273–5

magnetism activity: card sort sets 209–15; elaboration phase 206–7; engagement phase 203–4; evaluation phase 207–8; explanation phase 204, 206; exploration phase 204; implementation tools 209–16; sample chart 216
Magnusson, S. *et al.* 51
materials management 110–11
Mcinerney, D., & Mcinerney, V. 22
mealworms insect study *see* insect study activity
metacognitive questions 151
methods students' prior ideas *see* assessment activities
microteaching 94, 178
MO-STEP (Missouri Standards for Teacher Education Programs) 40–1
moon investigation activity: content standards 297; implementation tools 296–300; learning goals 287; overview 287–8; practical advice 295–6; procedure 288–95
Moore, J. J., & Watson, S. B. 71
motivation 18, 66, 68, 118, 182, 297
Musikul, K. 60

nature of science: card sort 141; creativity 70; and curriculum design 44; developing appropriate views of the 110; misconceptions about the 69; preservice teachers views of 70, 86; role of observation and inference 198, 202; tentative nature of scientific ideas *198*
nature of science knowledge, diagnostic assessment 141
nature of science objectives: electric circuits study 256; floating and sinking activity 219; human body activity 248; insect study activity 234; light and shadows activity 271; magnetism activity 202; moon investigation activity 288
NCATE (National Council for the Accreditation of Teacher Education) 39, 47
"need to know" 106, 112, 129, 257
norm-referenced scoring approach 152, 154
NSES (National Science Education Standards) 39

observation and inference, role of 198, 202

orient–enhance–synthesize model 87, 132, 256–7
orientations 3–5, 12, 50–5, 57, 59–63, 67–9, 78, 87, 92, 106, 108, 138, 140–1, 151, 204, 206; activity-driven 51, 53, 62; Anderson and Smith's list 51; card sort task 52–3, 60, 60–2; changing incoming 68–9; concept analysis 51, 68; diagnostic assessment 141; pedagogy-driven 53; reflection *see* reflection orientation; teacher inquiry 54; to teaching science teachers 51–4; topics 53
Osborne, R. *et al.* 252

partners model 181
PCK (pedagogical content knowledge): for assessment 4; confidence to teach science 3, 81; for curriculum 4; inputs 3; for instructional strategies 4, 233, 264; for learners 2, 4; orientations 51–4; refinement and revision of Shulman's notion of 2–3; requirements for development 65; for teaching science 4–5, 12, 69, 103–4, 125–6, 135, 196, 270; for teaching teachers 4, 5, 12, 196
pedagogical content knowledge *see* PCK
pedagogical knowledge objectives: floating and sinking activity 219; human body activity 248; insect study activity 234; light and shadows activity 271; magnetism activity 202; moon investigation activity 288
pedagogical principles: in electric currents activity 259; in floating and sinking activity 221, 223–4, 226–8; in human body activity 249
pedagogy-driven orientation 53
peer review 142, 208
personal constructivism 18–20, 22
Piaget, J. 19–20
planning science lessons/units *see* lesson planning assignment
planning the methods course *see* course planning
policy context, goals and standards 38–40
predict–observe–explain (POE) 87–8, 111, 118–19
principles of effective assessment *see* assessment principles
processing: deep vs surface 28; influences on 28
programme context: elementary teacher education 40, 40–3; field experience and partnerships 46; science courses 44–5

question-raising approach 126

raising hands 17, 109–11
reading reaction sheets 142, 148
Reading Recovery programme 180
reflection: and action research 184; exercises 188–90; foci for informal activities 183
reflection orientation 51, 54, 55, 101, 117, 221; case-based pedagogy 55–6; contexts 55; field experience 56–7; grounding 54; learning role 57–8; reading reactions 58; and the science methods course 54–9
resources for teaching elementary methods: case-based 55–6, 179; curriculum materials 94; facilities and equipment 95–7; non-text materials 92–3; online materials 92; planning matrix 100; the textbook 91–2
role play 267–8, 280, 282–4
rote learning 25, 27–8, 69, 105
rubrics and scoring tools 152–6, 159–60, 165, 167, 237, 240, 242–4, 253

safety and ethics, in course planning 97
schema theory 13–16
Schoon, K. J. and Boone, W. J. 71
science autobiography 141
science autobiography assignment 76
science classroom talk 274
science content: elementary majors' courses vs non-majors' courses 71; gaining confidence with 275; opportunities for emphasis 71; standards 206; teaching 135
science content knowledge, building 287
science content knowledge objectives: electric circuits study 256; floating and sinking activity 219; human body activity 248; insect study activity 234; light and shadows

science content knowledge objectives *continued*
  activity 271; magnetism activity 202; moon investigation activity 288
science content standards 206
science teacher educator PCK for teaching prospective teachers 5
science teaching notebooks 61, 236
science teaching standards 39
scoring tools and rubrics 139, 152–6, 159–60, 165, 167, 237, 240, 242–4, 253
seamless (embedded) assessments 149, 293
seamless assessment activity *see* moon investigation activity
self-assessment 41, 142, 165, 189
Shulman, L. S. 2, 199
situated learning 172–4, 178
Smith, D. 180
social cognitive learning theory 17–18
social constructivism 18, 20–2
Southerland, S. A., & Gess-Newsome, J. 72
stakeholders 37–8, 46, 173, 175, 185
standards, professional 39–40, 47, 107, 206
subject matter knowledge 3, 5, 69–70; the nature of science 70; science content 70–1
summative assessment 4, 113, 130, 134, 139–40, 143–6, 151–2, 206, 295
summative assessment assignment and scoring tools: science instructional materials analysis example 162–5; science teaching portfolio example 165–7
summative assessment techniques 143–4
surface processing 28
syllabus: component recommendations 98; *see also* course planning

TEAC (Teacher Accreditation Council) 39, 47–8
teacher-as-researcher 54, 62, 184
teacher certification 4, 37, 40, 42
teacher inquiry orientation 54
teacher knowledge: case knowledge 195; components 52; of context 3, 5; of learners 73–4; of nature of science 70; propositional 195; of science concepts 70–1; for science teacher educators 3

teacher learning 4, 172, 176, 181, 185
teacher licensing 40
teacher preparation, essential components 127
teacher preparation programmes, criticisms 42
teachers as learners 52, 66–7
teaching and learning cycle 42–3
teaching and learning cycle framework 43
teaching models: 5E model 130–2 (*see also* magnetism activity); conceptual change model 132; generative learning model 133; interactive approach 126–30; orient–enhance–synthesize model 87, 132, 256–7; overview 125–6; predict–observe–explain model 87–8, 111, 118–19
the textbook 93
thinking like a teacher 43, 58, 71, 142, 177, 184, 188, 221
Thomas, J. A. *et al.* 252
tools for teaching 60, 100, 162, 182, 187, 209, 229, 238, 265, 276, 284, 296; Course Planning Matrix 100; Field Experience Feedback 187, 187–90; Orientations to Teaching Teachers Card Sort Task 60, 60–2; Reflection on the Field Experience 182; Summative Assessment Assignment and Scoring Tool 162, 162–7
topics orientation 53, 82
tradebooks 144, 278–80, 282, 284, 301

videocases 55–8, 123, 136, 175, 179
Views of Nature of Science (VNOS) 78, 110, 141, 235
virtual field experience 178–9
Vygotsky, L. S. 20–2

wait time 89, 110–11, 115–16
water cycle activity: implementation tools 284–5; overview 278–81; practical advice 284; role play 282–4; think–pair–share 281–2
The Water's Journey (Schmid) 279
weekly plans 87–91, 100
weekly writing themes 298–9

ZPD (Zone of Proximal Development) 20–1